Proceedings of the
Steklov Institute of Mathematics
Volume 206 Issue 5 of 6 1995

The 21st Hilbert Problem for Linear Fuchsian Systems

A. A. Bolibrukh

A Translation of

ТРУДЫ
МАТЕМАТИЧЕСКОГО ИНСТИТУТА
имени В. А. СТЕКЛОВА

ТОМ 206 1994

ABSTRACT: The author's negative solution of Hilbert's twenty-first problem for linear Fuchsian systems of differential equations is presented. Methods developed by the author in solving this problem are then applied to the study of scalar Fuchsian equations and systems with regular singular points on the Riemann sphere. The monograph is intended for researchers and post-graduates specializing in the area of the analytic theory of differential equations and mathematical physics.

1991 *Mathematics Subject Classification.* Primary 34-02, 34A20, 34A30; Secondary 32S40, 32L05, 30F30.

РОССИЙСКАЯ АКАДЕМИЯ НАУК

ТРУДЫ

МАТЕМАТИЧЕСКОГО ИНСТИТУТА
имени В. А. СТЕКЛОВА

206

А. А. Болибрух

21-Я ПРОБЛЕМА ГИЛЬБЕРТА ДЛЯ ЛИНЕЙНЫХ ФУКСОВЫХ СИСТЕМ

Москва «Наука» 1994

Correspondence between *Trudy Mat. Inst. Steklov* and *Proc. Steklov Inst. Math.*

Russian imprint	Vol.	English imprint	Russian imprint	Vol.	English Imprint	Issue	Russian imprint	Vol.	English Imprint	Issue	Russian imprint	Vol.	English Imprint	Issue
1966	74	1967	1971	109	1974		1979	144	1980	2	1988	179	1989	2
1965	75	1967	1970	110	1972		1980	145	1981	1	1987	180	1989	3
1965	76	1967	1970	111	1972		1978	146	1980	3	1988	181	1989	4
1965	77	1967	1971	112	1973		1980	147	1981	2	1988	182	1990	1
1965	78	1967	1970	113	1972		1978	148	1980	4	1990	183	1991	4
1965	79	1966	1970	114	1974		1978	149	1981	3	1990	184	1991	2
1965	80	1968	1971	115	1974		1979	150	1981	4	1988	185	1990	2
1966	81	1968	1971	116	1973		1980	151	1982	2	1989	186	1991	1
1966	82	1967	1972	117	1974		1980	152	1982	3	1989	187	1990	3
1965	83	1967	1972	118	1976		1981	153	1982	4	1990	188	1991	3
1965	84	1968	1973	119	1976		1983	154	1984	4	1989	189	1990	4
1966	85	1967	1974	120	1976		1981	155	1983	1	1989	190	1992	1
1965	86	1967	1972	121	1974		1980	156	1983	2	1989	191	1992	2
1966	87	1967	1973	122	1975		1981	157	1983	3	1990	192	1992	3
1967	88	1969	1973	123	1975		1981	158	1983	4	1992	193	1993	3
1967	89	1968	1976	124	1978		1983	159	1984	2	1992	194	1993	4
1967	90	1969	1973	125	1975		1982	160	1984	1	1994	195	1995	2
1967	91	1969	1973	126	1975		1983	161	1984	3	1991	196	1992	4
1966	92	1968	1975	127	1977		1983	162	1985	1	1991	197	1993	1
1967	93	1970	1972	128	1974		1984	163	1985	4	1992	198	1994	1
1968	94	1969	1973	129	1976		1983	164	1985	2	1993	199	1995	1
1968	95	1971	1978	130	1979	4	1984	165	1985	3	1991	200	1993	2
1968	96	1970	1974	131	1975		1984	166	1986	1	1992	201	1994	2
1968	97	1969	1973	132	1975		1985	167	1986	2	1993	202	1994	4
1968	98	1971	1973	133	1977		1984	168	1986	3	1994	203	1995	3
1967	99	1968	1975	134	1977		1985	169	1986	4	1993	204	1994	3
1971	100	1974	1975	135	1978		1984	170	1987	1	1994	205	1995	4
1972	101	1975	1975	136	1978		1985	171	1987	2	1994	206	1995	5
1967	102	1970	1976	137	1978		1985	172	1987	3	1994	207	1995	6
1968	103	1970	1975	138	1977		1986	173	1987	4				
1968	104	1971	1980	139	1982	1	1986	174	1988	1				
1969	105	1971	1976	140	1979	1	1986	175	1988	2				
1969	106	1972	1976	141	1979	2	1987	176	1988	3				
1971	107	1976	1976	142	1979	3	1986	177	1988	4				
1968	108	1971	1977	143	1980	1	1986	178	1989	1				

This nonjournal serial does not always appear in strict sequential volume order in the original. Accordingly, the translation will occasionally reflect any gaps resulting thereby, as happens for example with the English imprints of 1984, 1985, and 1990.

CONTENTS

RUSSIAN CONTENTS*

*The American Mathematical Society scheme for transliteration of Cyrillic may be found at the end of index issues of *Mathematical Reviews.*

Introduction

1. The Riemann–Hilbert problem is connected with a certain special class of linear differential equations on the Riemann sphere CP^1: the class of so-called Fuchsian systems.

We consider a system of linear differential equations

$$df = \omega f \tag{1}$$

on the Riemann sphere, where $\omega = \|\omega_{ij}\|$, $1 \le i, j \le p$, is a matrix of differential 1-forms holomorphic on CP^1 with the exception of a finite set $\mathscr{D} = \{a_1, \dots, a_n\}$ of singular points. A point a_i is called a Fuchsian point of the system (1) if the form ω has a first-order pole at this point. Denote by $B^i = \operatorname{res}_{a_i} \omega$ the residues of ω at the points a_i. If the point ∞ is not among the singular points of a Fuchsian system (1), then, by the residue theorem,

$$\sum_{i=1}^{n} B^i = 0, \tag{2}$$

and hence the differential matrix form

$$\delta = \omega - \sum_{i=1}^{n} B^i \frac{dz}{z - a_i}$$

is holomorphic on the whole Riemann sphere. Indeed, the form δ is holomorphic on the complex plane C by construction, and in the coordinate $\zeta = 1/z$ it has in view of (2) the appearance

$$\delta(\zeta) = \omega(\zeta) + \sum_{i=1}^{n} B^i \frac{d\zeta}{\zeta} - \sum_{i=1}^{n} a_i B^i \frac{d\zeta}{a_i\zeta - 1} = \omega(\zeta) - \sum_{i=1}^{n} \frac{a_i}{a_i\zeta - 1} B^i d\zeta$$

in a neighborhood of ∞ $(\zeta = 0)$, that is, δ is holomorphic also at $\zeta = 0$. Thus, the differential form δ is holomorphic on the whole Riemann sphere, and hence $\delta \equiv 0$.

Thus, if ∞ is not among the singular points of the Fuchsian system (1), then in the coordinate z of the complex plane this system can be written as

$$\frac{df}{dz} = \left(\sum_{i=1}^{n} B^i \frac{1}{z - a_i} \right) f. \tag{3}$$

It can also be verified by direct computation that a conformal transformation $w = (az + b)/(cz + d)$ of the Riemann sphere carries the Fuchsian system (3) into a Fuchsian system with the same residues at the singular points $a_i' = (aa_i + b)/(ca_i + d)$.

At a point $z_0 \in CP^1 \setminus \mathscr{D}$ we consider the germ $\widehat{T}(z_0)$ of a fundamental matrix $T(z)$ of the solution space X of the system (1) (the columns of $T(z)$ form a basis in X). Denote by $\widehat{T}'(z_0)$ the analytic continuation of $\widehat{T}(z_0)$ around a loop γ lying in $CP^1 \setminus \mathscr{D}$. The germ $\widehat{T}'(z_0)$ is also the germ of some fundamental matrix of X; therefore,

$$\widehat{T}'(z_0)G = \widehat{T}(z_0), \qquad G \in GL(p; C).$$

The correspondence $\gamma \mapsto G$ depends only on the homotopy class $[\gamma]$ of the loop γ, and it yields a homomorphism

$$\chi : \pi_1(CP^1 \setminus \mathscr{D}, z_0) \to GL(p; C) \tag{4}$$

of the fundamental group of the space $CP^1 \setminus \mathscr{D}$ into the group of nonsingular complex $p \times p$ matrices. (Properly speaking, it is a homomorphism of the covering transformation group of the universal covering $\widetilde{CP^1 \setminus \mathscr{D}}$ of the space $CP^1 \setminus \mathscr{D}$ into $GL(p, C)$, but as a rule it is more convenient to work with a more concretely given object—the fundamental group, which can be identified in a familiar way with the covering transformation group; see §4 in Chapter I.) The homomorphism (4) is called the monodromy representation or simply the monodromy of the system (1), and the group $\operatorname{Im}\chi$ is called the monodromy group of this system. When the point z_0 is replaced by a point z_0' or when the fundamental matrix $T(z)$ is changed, the monodromy matrices G pass into $S^{-1}GS$, where S is a nonsingular constant matrix. The monodromy of the system (1) is thus determined up to equivalence (see §1 in Chapter I).

The definition of the monodromy representation is analogous for Fuchsian linear differential equations

$$u^{(p)} + q_1(z)u^{(p-1)} + \cdots + q_p(z)u = 0 \tag{5}$$

(see §1 in Chapter III for the definition of a Fuchsian equation). Here in place of the matrix $T(z)$ we should consider the matrix with ith column of the form

$$\left(u_i, \frac{du_i}{dz}, \ldots, \frac{d^{p-1}u_i}{dz^{p-1}}\right)^T,$$

where $u_1, \ldots, u_p$ is a basis in the solution space of equation (5), and T denotes transposition.

All singular points are regular singular points for a Fuchsian system (3). This means that, as a singular point a_i is approached along any sectorial neighborhood different from C and with vertex at a_i, any solution f increases no more rapidly than some power of the distance $|z - a_i|$ to this point.

The class of systems (1) with regular singular points contains the class of Fuchsian systems, but is not exhausted by it (see [15], [25]).

In contrast to the case for systems, the concept of being Fuchsian is equivalent to the concept of regularity for equations (5) (see [7], [15]).

2. The problem of reconstructing a Fuchsian equation from its monodromy (4) was first mentioned by Riemann in a note at the end of the 1850's (see [23]). In 1900 Hilbert included it as number XXI among his "Mathematische Probleme" [21]:

"*zu zeigen, daß es stets eine lineare Differentialgleichung der Fuchsschen Klasse mit gegebenen singulären Stellen und einer gegebenen Monodromiegruppe giebt.*"

In connection with Fuchsian systems it has become traditional in the literature to refer to this problem as the Riemann–Hilbert (RH) problem. (The same name is also used for another problem, which is not considered here; see, for example, [9].) By a conformal transformation of CP^1 we can always ensure that ∞ is not among the singular points of (1), so the RH problem can be formulated as follows:

Given a representation (4), *show that there is always a system* (3), (2) *with the monodromy* (4).

It was believed for a long time that Plemelj had completely solved the RH problem in 1908 [52]. However, gaps in his proof were discovered at the beginning of the 1980's (see [2], [46]). The method of solution proposed by Plemelj consisted in reducing the RH problem to the so-called homogeneous Hilbert boundary value problem in the theory of singular integral equations. A thorough investigation of the latter problem is contained in the books of Muskhelishvili [19] and Vekua [9]. This reduction is implemented as follows.

Suppose that all the points $a_1, \dots, a_n$ lie in the finite complex plane, and let them be joined by a simple closed contour L, as in the following diagram:

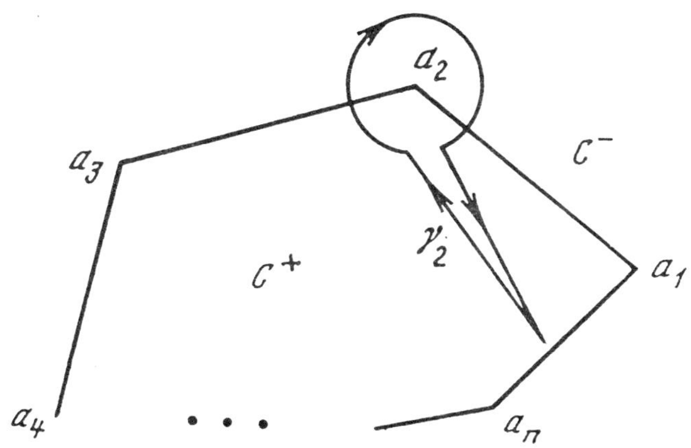

Let $g(t)$ be a piecewise constant nonsingular matrix-valued function on L given by

$$g(t) = G_i \cdot \dots \cdot G_1, \qquad t \in [a_i; a_{i+1}),$$

where G_i is the monodromy matrix (4) corresponding to a circuit of the point a_i along a "small" loop. Denote by C^+ the domain bounded by L in the finite complex plane, and by C^- the complement of $\overline{C}^+$ in CP^1.

Consider the following problem: find all the pairs of vector-valued functions $\varphi^+ = (\varphi^+_{(1)}, \dots, \varphi^+_{(p)})$ and $\varphi^- = (\varphi^-_{(1)}, \dots, \varphi^-_{(p)})$ such that:

1) $\varphi^+(z)$ is holomorphic in C^+, and $\varphi^-(z)$ is holomorphic in C^- and has power growth at ∞;

2) $\varphi^+(z)$ and $\varphi^-(z)$ are continuous up to the contour L with the exception of the points $a_1, \dots, a_n$, and they are connected on (a_i, a_{i+1}) by the relation

$$\varphi^+(t)g(t) = \varphi^-(t);$$

3) $\varphi^{\pm}(z)(z-a_i)^{\varepsilon}$ tends to zero for some $0 \le \varepsilon < 1$ as z approaches a_i in the respective domains C^+ and C^-.

This problem is reduced to the problem with a continuous function $g(t)$ (Plemelj and Vekua realize such a reduction in different ways), and this is then solved by methods in the theory of singular integral equations. Here it turns out that there is always a system of solutions $\varphi_1^{\pm}(z), \dots, \varphi_p^{\pm}(z)$ for which the following conditions hold:

a) the determinant of the matrix T with rows made up of the vector-valued functions $\varphi_1^{\pm}(z), \dots, \varphi_p^{\pm}(z)$ is nonzero at all points of the complex plane C except the points in $\mathscr{D}$;

b) the matrix $z^S \cdot T(z)$, where S is an integer diagonal matrix, is holomorphically invertible at ∞.

A function $\varphi_l^+(z)$ in this system of solutions can be continued analytically to any point in $C \setminus \mathscr{D}$ along any path disjoint from the set $\mathscr{D}$ of singular points. It follows from the property 2) that when φ_l^+ is continued along a "small" loop going around a_i, it passes into $\varphi_l^+ G_i^{-1}$ ($i=2$ in the diagram); therefore, the same is true for the matrix $T(z)$. Thus, T passes into a matrix $\widetilde{T}$ under analytic continuation along a loop γ beginning and ending at z_0, where

$$\widetilde{T}G = T, \qquad G = \chi([\gamma]).$$

The same is true for the matrix $T' = (z-a_1)^S T$, which is holomorphically invertible at ∞ according to property b) of the system of solutions.

This and property a) imply that the matrix form

$$\omega = dT'(T')^{-1} \tag{6}$$

is single-valued on CP^1 and holomorphic away from the singular points $a_1, \dots, a_n$ in $\mathscr{D}$. The system (3) with matrix form (6) has the given monodromy (4), and the points $a_1, \dots, a_n$ are regular singular points for it.

Plemelj next employs a procedure allowing him to pass from the system constructed to another system that has the same monodromy and the same singular points but is now Fuchsian at all but perhaps one of the points. There are no objections to this part of his proof. However, for the general case he does not give a rigorous proof of his claim that the system can be reduced to a Fuchsian system also at the last point, though his argument does work if one of the monodromy matrices G_i is diagonalizable (see [2]).

Thus, Plemelj proved that the RH problem is solvable in the case when one of the monodromy matrices is diagonalizable, say the matrix G_i corresponding to a "small" loop around a_i.

Plemelj also first solved the analogue of the RH problem in the class of systems with regular singular points.

After the appearance of Plemelj's result [52] the theme of work connected with the RH problem shifted mainly toward the effective construction of the matrices of a Fuchsian system from given monodromy matrices $G_1, \dots, G_n$. At the end of the 1920's Lappo-Danilevskiĭ used his method of analytic functions of matrices to represent the solutions of a Fuchsian system and the monodromy matrices $G_1, \dots, G_n$ as convergent series in the matrices of coefficients in the system (see [18]). In this case the effective solution of the RH problem was reduced to inversion of the series obtained and to an investigation of the question of convergence, which was answered positively in [18] for matrices $G_1, \dots, G_n$ close to the identity.

Lappo-Danilevskiĭ thereby proved that the RH problem is solvable for representations (4) whose monodromy matrices corresponding to "small" loops around the points a_i are close to the identity.

In 1956 Krylov [17] proved by constructing an effective solution that the RH problem is solvable for a representation (4) of dimension $p = 2$ when there are three singular points.

The analogous problem for four singular points was treated by Erugin [14], who established, in particular, that it is connected with the Painlevé equation.

The problem of effectively constructing an equation of Fuchsian type is closely connected with the problem of computing accessory coefficients, which is extremely important for applications but will not be considered here. See [14], [10], and [15] about this question.

A new stage in the study of the RH problem was opened in 1957 by Röhrl [53], who was the first to apply methods in the theory of bundles to its solution. (Considerations of this kind actually go back to Birkhoff [38], who reproved Plemelj's result, but at that time there was no adequate geometric language for an appropriate description.)

From the representation (4) Röhrl constructed a principal bundle F' on $CP^1 \setminus \mathscr{D}$ with structure group $GL(p; C)$. The bundle turns out to be holomorphically trivial (see [25]); therefore, the form ω in (6) constructed from the holomorphic trivialization $T'(z)$ of this bundle determines a system (1) with the given singular points and the given monodromy. Röhrl then extended the bundle F' to the whole Riemann sphere CP^1 with the help of the section $T'(z)$. The extended bundle always has a meromorphic section that is holomorphically invertible away from the points in $\mathscr{D}$. The system (1) constructed from this section has the given monodromy, and the points $a_1, \dots, a_n$ are regular singular points for it.

Thus, Röhrl reproved Plemelj's results, along with the solvability of the RH problem on a noncompact Riemann surface. Moreover, in [53] and [54] he proved that the analogue of the RH problem for an arbitrary Riemann surface is solvable in the class of systems with regular singular points (it is true that additional "false" singularities not contributing to the monodromy appear for the system constructed).

A "topological" approach to the RH problem permits one to appraise the set of representations for which the modified RH problem is solvable (Il′yashenko [2]): "The space $GL(p; C))^{n-1}$ contains a countable union of proper analytic submanifolds such that any collection $G_1, \dots, G_{n-1}$ in its complement has the following property. For each sequence $(z - a_1)^{E_1}, \dots, (z - a_n)^{E_n}$ of Fuchsian principal parts with

$$\exp(2\pi i E_j) = G_j, \qquad j = 1, \dots, n,$$

$$G_n = (G_1 \cdot \dots \cdot G_{n-1})^{-1}, \qquad \sum_{i=1}^{n} \operatorname{tr} E_j = 0$$

there exists a Fuchsian system whose fundamental matrix of solutions has Fuchsian parts in the given sequence."

In 1976 and 1979 the papers [40] and [41] of Dekkers appeared, which lead to the solvability of the RH problem for any collection of points $a_1, \dots, a_n$ and any representation (4) of dimension $p = 2$. The method of solution consists in correcting a system (1) with regular singular points and with the given monodromy to form a Fuchsian system with the help of gauge transformations

$$g = \Gamma(z) f \tag{7}$$

under which the system (1) passes into the system $dg = \omega' g$ with form

$$\omega' = d\Gamma \cdot \Gamma^{-1} + \Gamma \omega \Gamma^{-1}. \tag{8}$$

If the matrix-valued function $\Gamma(z)$ is meromorphic on CP^1 and is holomorphically invertible away from the points in $\mathscr{D}$, then the system (1), (8) is again a system with regular singular points $a_1, \dots, a_n$ and the same monodromy. Using a result of Deligne [42] about the form of the coefficient matrix of a system with regular singular points, Dekkers proved the existence of a matrix $\Gamma(z)$ reducing ω to the matrix ω' in (8) which has singularities of the simple pole type at the points in $\mathscr{D}$.

Different aspects of the RH problem and problems connected with it were also treated in [13], [24], [30], [35], [39], [42], and [45].

3. In the first two chapters we carry out a thorough investigation of the RH problem for a representation (4) of dimension $p = 3$. Chapter I is mainly of a preparatory nature. In §1 we give the definitions of a regular singular point and of the rate of growth of a solution at such a point—the so-called normalization. In §2 we present results, due to Levelt [49] and Gantmakher [11], on the structure of solution spaces of Fuchsian systems and systems with regular singular points.

In §3 we formulate some facts from the theory of systems with reducible monodromy, and describe technical devices to be used in what follows. The basic method in this monograph is the method of gauge transformations (7), which amounts to passing from the system (1) to the system with the form (8). This changes the character of the singularities of the system at the points in $\mathscr{D}$. In some cases it is possible to trace the changes with the help of (8) (see [41]), but technically this is fairly complicated to do. For our purposes it is much more effective to trace the change in the structure of the space of solutions (1.2.6) under the transformations (7), and this is what we do.

Section 4 is central in Chapter I. Here we describe the class of holomorphic bundles on the Riemann sphere that are extensions of the Röhrl bundle F' and have the property that the RH problem is solvable if and only if one of the constructed bundles is holomorphically trivial (Theorem 1.4.1). Here necessary and sufficient conditions for the positive solvability of the RH problem are obtained for a special class of reducible representations (4) in terms of the partial indices of the canonical extension of the Röhrl bundle F' (Theorem 1.4.2).

As an illustration of the use of the methods developed in §§1–4 we reprove the above-cited results of Plemelj (Proposition 1.5.2) and Dekkers (Proposition 1.5.3) in §5.

The main results in Chapter II are the following (see [3]–[5]).

THEOREM 2.1.1. *For any irreducible representation* (4) *of dimension* $p = 3$ *the RH problem has a positive solution.*

THEOREM 2.3.3. *For any* $n > 3$, *any points* $a_1, \ldots, a_n$, *and any* $p \geq 3$ *there exists a representation* (4) *not realizable by a Fuchsian system.*

Theorem 2.3.3 means that the RH problem has a negative solution in the general case.

Theorem 2.1.1 is proved in §1 of Chapter II. Theorem 2.3.3 is proved in §3 on the basis of results in §2. Here all the representations of dimension $p = 3$ that are not realized by Fuchsian systems are described (Theorem 2.3.1). All such representations (4) are reducible, and each of the generators of the group $\operatorname{Im} \chi$ of such a monodromy (4) is reduced to a Jordan normal form consisting of a single block. An obstacle to the positive solvability of the RH problem in this case is the so-called Fuchsian weight of the corresponding two-dimensional subrepresentation (or factor representation). The concept of a Fuchsian weight is introduced and studied in §2. For an arbitrary Fuchsian system of two equations the Fuchsian weight is defined as the sum over all singular points of the moduli of the differences of the rates of growth (the normalizations) of the basis solutions of the system (see Definition 2.2.1). The Fuchsian weight of the representation (4) is the minimal value of the Fuchsian weights of all possible Fuchsian systems with the given monodromy (Definition 2.2.2).

In §6 the Fuchsian weight of a two-dimensional representation (4) is interpreted in terms of the partial indices of the canonical extension F^0 of the Röhrl bundle F' (Theorem 2.1.6). It turns out that the holomorphic type of the bundle F^0 is completely determined by its first Chern class (a topological invariant) and by the Fuchsian weight of the representation (4) from which F^0 is constructed (an analytical invariant).

In §5 we prove the instability of the Fuchsian weight (Proposition 2.5.2) with respect to displacements of the points $a_1, \ldots, a_n$. It turns out that under preservation of the monodromy matrices there exists an arbitrarily small shift of one of the singular points for which the Fuchsian weight of the representation drops to 0 or 1. This implies that all the examples of a negative solution of the RH problem with $p = 3$ bear an unstable character, in the sense that there exists an arbitrarily small displacement of one of the singular points under which the RH problem with the same monodromy matrices becomes solvable in the positive sense.

The last fact explains why a representation (4) of dimensional $p = 3$ for which the RH problem has a negative solution is presented in §4 not explicitly but as the monodromy of a system (2.4.1) with four regular singular points. The goal of this section is to analyze a concrete example of a negative solution of the RH problem (Theorem 2.4.1), and the section thus contains a minimal number of references to theorems of general character from §§1–3. The corresponding assertions are reproved in §4 in a simplified version adapted to this concrete situation.

As is known (see, for example, [51]), the problem of constructing a Fuchsian linear differential equation of order p with a given monodromy and given singular points has a negative solution in the general case for $p > 1$, because such an equation contains fewer parameters than the set of equivalence classes of representations (4) (this count of parameters goes back to Poincaré [22]; see §1 in Chapter III). Therefore, in the construction of the Fuchsian equation additional singular points of the coefficients arise that do not contribute to the monodromy and are called "false" singular points. The first section of Chapter III gives an upper estimate for the number of such points in terms of the dimension p of the representation and the partial indices of the canonical extension of the Röhrl bundle F' (Theorem 3.1.1).

From any equation that is Fuchsian in a neighborhood of the point a_i it is possible to construct a Fuchsian system by means of a special substitution, but this construction bears an essentially local character [16]. The following assertion is proved in §2.

THEOREM 3.2.1. *From any Fuchsian equation on the Riemann sphere it is possible to construct a Fuchsian system* (3) *with the same singular points and the same monodromy.*

Theorem 3.2.1 answers a question posed by Erugin [14]. It enables us also to present examples of representations (4) that trivially cannot be realized as monodromy representations of Fuchsian equations without additional false singular points (Proposition 3.2.1, Example 3.2.1).

Chapter IV is devoted to continuous representations. In §2 of this chapter we introduce an important invariant: the maximal Fuchsian weight of a representation. We consider the family $\{F^{\varphi}\}$ of bundles constructed in §4 of Chapter I from a representation (4), along with the splitting type $c_1^{\varphi}, \dots, \varphi_p^{\varphi}$ of each of those bundles. The number

$$\gamma_m(\chi) = \sup_{\varphi} \sum_{i=1}^{p} (c_1^{\varphi} - c_i^{\varphi}) \tag{9}$$

is called the maximal Fuchsian weight of the representation (4) (see Definition 4.2.1). It is proved in Theorem 4.2.1 that this number is finite if and only if the representation is irreducible. In this case

$$\gamma_m(\chi) \leq (n-2)p(p-1)/2. \tag{10}$$

The main result in §3 of Chapter IV (and in fact in the whole chapter) is

THEOREM 4.3.1. *The RH problem has a positive solution for any irreducible representation* (4).

(See [7]; an analogous result was obtained independently in [48].) The proof of this theorem is based on the finiteness (10) of the maximal Fuchsian weight of an irreducible representation and uses the technical Lemma 4.1.3 in §1. This theorem is a generalization of the result of (but not of the proof of) Theorem 2.1.1 to the case of a representation of arbitrary dimension. (The proof of Theorem 2.1.1, like those of the other main results in the first three chapters, has a constructive character and can be algorithmized in principle.)

In §4 of Chapter IV we find the exact value for the minimal possible number m_0 of additional false singular points arising in the construction of a scalar Fuchsian differential equation with given irreducible monodromy (4). According to Theorem 4.4.1, this number is equal to

$$m_0 = (n-2)p(p-1)/2 - \gamma_m(\chi). \tag{11}$$

(The fact that m_0 does not exceed the right-hand side of (11) actually follows from the inequality (3.1.29) and Remark 3.1.2 in §1 of Chapter III; Theorem 4.4.1 asserts that equality holds.)

We say that a family of bundles on the Riemann sphere has the finiteness property if its maximal weight (defined like the maximal Fuchsian weight of the family $\{F^{\varphi}\}$ of bundles constructed above) is finite. The property of finiteness of some special family of bundles plays a principal role in the proof of the main result in §5 of Chapter IV. Here we consider a system of equations

$$z\frac{df}{dz} = A(z)f \tag{12}$$

with coefficient matrix $A(z)$ that is meromorphic at ∞, having a pole of order r there. (We remark that for $r > 0$ the point ∞ is in general an irregular singular point for the system (12).) If by means of a transformation (7) that is holomorphically invertible in a neighborhood of ∞ the system (12) can be reduced to a system of the form (12) whose coefficient matrix is a polynomial in z, then we say that the original system can be reduced to Birkhoff standard form by means of an analytic representation (7).

Birkhoff [37] claimed that any system (12) can be reduced by an analytic transformation to Birkhoff standard form, but Gantmakher [11] constructed a counterexample. It turned out that Birkhoff's proof works only for the case when the monodromy matrix of (12) is diagonalizable.

We say that the system (12) is reducible if it can be reduced by some analytic transformation (7) to a system with a subsystem of smaller dimension. For the case $p = 2$ Jurkat, Lutz, and Peyerimhoff [44] proved that each irreducible system (12) can be reduced to Birkhoff standard form by an analytic transformation, and Balser [32] proved the same thing for $p = 3$. In Theorem 4.5.1 of §5 we prove that the analogue is true for any p.

In §6 of Chapter IV the results in §6 of Chapter II are carried over to the case of an arbitrary number of equations. Here we establish that the asymptotic behavior of the solutions of a Fuchsian system at singular points is connected with the splitting type of the canonical extension of the bundle constructed from the monodromy of this system, we show that in any holomorphic bundle on the Riemann sphere there exists an irreducible meromorphic connection with logarithmic singular points, and we find the minimal possible number n of such points (Corollary 4.6.3).

In §1 of Chapter V it is proved that any reducible representation (4) can be realized as the monodromy representation of some system with regular singular points and with coefficient matrix having the same reducibility character as the monodromy (4) (Corollary 5.1.1). Here it is also shown that the analogous assertion is not true for Fuchsian systems (Proposition 5.1.1).

In §2 we give a "minimal" (with respect to the dimension $p = 4$ and the number $n = 3$ of singular points) example, stable under displacement of singular points, of a representation (4) that cannot be realized as the monodromy representation of a Fuchsian system (Example 5.2.1).

A negative solution of the RH problem means that there are systems with regular singular points on the Riemann sphere that, in contrast to the local situation, are not globally meromorphically equivalent to any Fuchsian systems. It is proved in Theorem 5.3.3 that any system (1) with regular singular points is a subsystem (factor system) of some system with the same singular points that is meromorphically equivalent to a Fuchsian system. The latter means that any representation (4) is a subrepresentation (factor representation) of a representation for which the RH problem has a positive solution. Sufficient conditions for the realizability of a reducible representation (4) of a Fuchsian system are given in Theorems 5.3.1 and 5.3.2 and Propositions 5.3.1 and 5.3.2 in §3.

4. The first three chapters are a reworking of the author's doctoral dissertation. The monograph does not treat questions connected with estimates of the orders of poles of a system with regular singular points nor with estimates of the codimension of the representations not realizable by Fuchsian systems in the space of all representations (4) with given singular points. These questions are studied in [29] (where, in particular, it is shown in Theorem 5.6.1 that this codimension is equal to $(p-1)(2n-1)$ for sufficiently large p and n). In the same reference there is a thorough discussion of the various formulations of the RH problem.

I am deeply grateful to V. A. Golubeva and A. V. Chernavskiĭ, who stimulated me to work on this topic, and to M. M. Postnikov, whose seminar on algebraic topology and its applications included several discussions of the results presented here. I would like to express my special thanks to D. V. Anosov, who read through a large part of the monograph and suggested a simplified version of the proof of Lemma 1.3.4 and a coordinate-free way of describing the canonical extension of the bundle constructed from a representation (4) (in place of the coordinate description that I had used).

This work was begun at the Moscow Institute of Engineering Physics and completed at the Steklov Mathematical Institute of the Russian Academy of Sciences. I thank my colleagues at both institutes for their attention and support.

CHAPTER I

Main Definitions. The Method of Solution

1. Main definitions. Regular singular points. Normalizations

Let us consider a system

$$df = \omega f \tag{1.1}$$

of linear differential equations on the Riemann sphere CP^1. Here ω is a matrix-valued differential 1-form. In the coordinate z of the complex plane C this form can be written as

$$\omega = B(z)dz, \tag{1.2}$$

where $B(z)$ is a $p \times p$ matrix-valued function. We consider only systems (1.1) whose matrix forms ω are holomorphic on the whole Riemann sphere CP^1 with the exception of a finite set $\mathscr{D} = \{a_1, \dots, a_n\}$ of points. The set X of solutions of such a system is a p-dimensional linear space of vector-valued functions that are holomorphic on the universal covering

$$\widetilde{CP^1\backslash\mathscr{D}} \xrightarrow{\pi} CP^1\backslash\mathscr{D}.$$

We shall denote points in $\widetilde{CP^1 \backslash \mathscr{D}}$ by y, and the points $\pi(y)$ in $CP^1 \backslash \mathscr{D}$ corresponding to them by z.

We fix once and for all a point z_0 in the base $CP^1 \backslash \mathscr{D}$ and a point y_0 in the covering $\widetilde{CP^1 \backslash \mathscr{D}}$ with $\pi(y_0) = z_0$.

The fundamental group $\pi_1(CP^1 \backslash \mathscr{D}, z_0)$ is now identified with the covering transformation group Δ; therefore, we can define an action g^* of an element $g \in \pi_1(CP^1 \backslash \mathscr{D}, z_0)$ on a function $f(y) \in X$ by

$$g^*(f) = f \circ g^{-1}. \tag{1.3}$$

Let $T(y)$ be a fundamental matrix of the solution space X, with its columns $e_1(y), \dots, e_p(y)$ forming a basis in this space. The matrix $g^*(T)$ also is a fundamental matrix of X, and hence is connected with the original matrix $T(y)$ by a relation

$$g^*(T) = T \circ g^{-1} = T \cdot \chi(g), \tag{1.4}$$

with $\chi(g) \in GL(p; C)$.

The action (1.3) determines a matrix representation

$$\chi\colon \pi_1(CP^1\backslash\mathscr{D}, z_0) \to GL(p; C) \tag{1.5}$$

of the fundamental group of the complement of the set $\mathscr{D}$ of singular points of (1.1) on the Riemann sphere over the field C.

Indeed, for elements $g, h \in \pi_1(CP^1 \setminus \mathscr{D}, z_0)$ we get that

$$\begin{aligned} T \cdot \chi(gh) &= (gh)^*(T) = T \circ (gh)^{-1} = T \circ h^{-1} \circ g^{-1} \\ &= (T \cdot \chi(h)) \circ g^{-1} = (T \circ g^{-1}) \cdot \chi(h) = T \cdot \chi(g) \cdot \chi(h), \end{aligned}$$

and thus $\chi(gh) = \chi(g)\chi(h)$. (If instead of using (1.3) we had given the action g^* by $g^*(f) = f \circ g$, we would have obtained an antirepresentation).

The representation (1.5) is called the *monodromy representation* of the system (1.1), or simply the *monodromy*. If instead of the fundamental matrix $T(y)$ we consider another fundamental matrix $\widetilde{T}(y)$, then (1.5) is replaced by an equivalent representation $\widetilde{\chi}$. Indeed, the matrices T and $\widetilde{T}$ are connected by the relation $\widetilde{T} = TS$, $S \in GL(p; C)$, and so

$$\widetilde{T} \cdot \widetilde{\chi}(g) = g^*(\widetilde{T}) = g^*(TS) = TS \circ g^{-1} = (T \cdot g^{-1})S = T \cdot \chi(g)S = \widetilde{T}(S^{-1}\chi(g)S),$$

which implies that $\widetilde{\chi}(g) = S^{-1}\chi(g)S$. The monodromy changes in exactly the same way when z_0 is replaced by a point z_0' (here the character of the identification of the fundamental group with the covering transformation group changes).

Let O_i denote connected simply connected disjoint neighborhoods of the points a_i in CP^1. In each neighborhood $O_i \setminus a_i$ we fix a point z_i^0 and consider a loop δ_i, with initial point and terminal point at z_i^0, corresponding to a single clockwise circuit of a_i. In $CP^1 \setminus \mathscr{D}$ we fix paths γ_i going from z_0 to the points z_i^0 in such a way that the loop $\prod_{i=1}^n (\gamma_i \delta_i \gamma_i^{-1})$ is homotopic to zero (the Riemann circuit relation). We denote the homotopy class of this loop by g_i.

The choice of the path γ_i induces a homomorphism

$$\gamma_i^* : \pi_1(O_i \setminus a_i, z_i^0) \to \pi_1(CP^1 \setminus \mathscr{D}, z_0). \tag{1.6}$$

DEFINITION 1.1.1. The representation

$$\chi_i : \pi_1(O_i \setminus a_i, z_i^0) \to GL(p; C), \tag{1.7}$$

with $\chi_i = \chi \circ \gamma_i^*$, is called the *ith local representation constructed from the representation* (1.5).

The restriction of the universal covering space $\widetilde{CP^1 \setminus \mathscr{D}}$ to O_i is disconnected when $n \geq 3$. Denote by $\widetilde{O}_i$ the connected component in this restriction that contains the terminal point y_i^0 of the lifting of γ_i to $\widetilde{CP^1 \setminus \mathscr{D}}$ with initial point at y_0. The triple $(\widetilde{O}_i, \pi_i, O_i \setminus a_i)$ is the universal covering of the space $O_i \setminus a_i$.

We consider on $\widetilde{O}_i$ the function $\ln(y - a_i)$ defined as follows. For any $y \in \widetilde{O}_i$ let

$$\ln(y - a_i) = \int_\gamma \frac{d(z - a_i)}{z - a_i},$$

where $\gamma = \gamma_i \cdot \widetilde{\gamma}$, and $\widetilde{\gamma}$ is a path in $O_i \setminus a_i$ joining z_i^0 and $\pi(y)$ such that its lifting to $\widetilde{O}_i$ is a path from y_i^0 to y.

Everywhere in what follows, the function $(y - a_i)^E$ is understood to be the function

$$(y - a_i)^E = \exp(E \ln(y - a_i)).$$

From the definition of the element $g_i \in \pi_1(CP^1 \setminus \mathscr{D}, z_0)$ it follows that

$$(g_i^*)((y-a_i)^E) = (y-a_i)^E \exp 2\pi i E. \tag{1.8}$$

This follows from the fact that, by the definition of g_i and (1.3),

$$g_i^*(\ln(y-a_i)) = \ln(y-a_i) + 2\pi i. \tag{1.8$'$}$$

The fundamental group $\pi_1(CP^1 \setminus \mathscr{D}, z_0)$ can be regarded as a group with n generators $g_1, \dots, g_n$ and with the single relation $g_1 \cdots g_n = 1$. Denote by G_i and E_i the matrices

$$G_i = \chi(g_i), \qquad E_i = \frac{1}{2\pi i} \ln G_i, \qquad 0 \le \operatorname{Re} \rho_i^j < 1, \tag{1.9}$$

where χ is the representation (1.5), and ρ_i^j, $j = 1, \dots, p$, are the eigenvalues of E_i. (The condition $0 \le \operatorname{Re} \rho_i^j < 1$ fixes uniquely the choice of the matrix $\frac{1}{2\pi i} \ln G_i$, whose definition in the general case contains some arbitrariness.)

We consider a fundamental matrix $T(y)$ of the solution space X of the system (1.1) on the universal covering $\widetilde{O}_i$ and the representation (1.5) constructed from $T(y)$.

The following proposition holds (see [15], [25]).

Proposition 1.1.1. *The fundamental matrix* $T(y)$ *is representable in* $\widetilde{O}_i$ *as*

$$T(y) = M_i(z)(y-a_i)^{E_i}, \tag{1.10}$$

where $M_i(z)$ *is a single-valued matrix-valued function that is holomorphically invertible in* $O_i \setminus a_i$, *and* E_i *is defined in* (1.9).

Proof. From (1.4), (1.9), and (1.8) it follows that

$$T^{-1}(y)(g_i^*(T))(y) = G_i = (y-a_i)^{-E_i} g_i^*((y-a_i)^{E_i}); \tag{1.10$'$}$$

therefore, the matrix-valued function $M(z) = T(y) \cdot (y-a_i)^{-E_i}$ is single-valued on $O_i \setminus a_i$ (in such cases we write z in place of the arugment y, with the understanding that $z = \pi(y)$). The holomorphic invertibility of $M(z)$ follows from the fact that

$$\det T(y) \ne 0, \qquad \det(y-a_i)^{E_i} = (y-a_i)^{\operatorname{tr} E_i} \ne 0$$

for all $y \in \widetilde{O}_i$.

Before passing to a description of the character of the singularities of the system (1.1) at the points $a_1, \dots, a_n$ we consider some auxiliary definitions.

Definition 1.1.2. A subset $K \subset \widetilde{O}_i$ will be called a *sectorial neighborhood* if the set $\pi(K)$ lies interior to some sector not coinciding with C and with vertex at a_i, and it contains a_i in its closure.

Definition 1.1.3. We say that the *limit* of a function $f(y)$ at a_i is equal to b, and write

$$\operatorname*{Lim}_{y \to a_i} f(y) = b,$$

if

$$\lim_{\substack{\pi(y) \to a_i \\ y \in K}} f(y) = b$$

for any sectorial neighborhood $K \subset \widetilde{O}_i$.

EXAMPLE 1.1.1. It is not hard to verify that for any $\lambda \in R$ with $\lambda > 0$

$$\operatorname*{Lim}_{y\to a_i}(y - a_i)^\lambda \ln(y - a_i) = 0.$$

It is clear that for a single-valued function on the neighborhood O_i Definition 1.1.3 is equivalent to the usual definition of the limit of a function at a point.

DEFINITION 1.1.4. We say that a (scalar-, vector-, or matrix-valued) function $f(y)$ on $\widetilde{O}_i$ *has finite order at* a_i if there exists a number $\lambda \in R$ such that

$$\operatorname*{Lim}_{y\to a_i}(y - a_i)^\lambda f(y) = 0. \tag{1.11}$$

We consider again a singular point a_i of the system (1.1).

DEFINITION 1.1.5. A point a_i is called a *regular singular point* of (1.1) if all the solutions of this system have finite order there.

If a_i is a regular singular point for the system (1.1), then in the decomposition (1.10) for the fundamental matrix $T(y)$ of this system the matrix-valued function $M_i(z)$ is meromorphic at a_i. Indeed, $T(y)$ has finite order at a_i because its columns have that property; the same is true for the matrix-valued function $(y - a_i)^{E_i}$, since it suffices to take $\lambda = 1$ as λ in (1.11) in view of Example 1.1.1 and the condition (1.9). Therefore, the matrix $M_i(z)$ also has finite order at a_i, and this is equivalent to $M_i(z)$ being meromorphic at a_i in view of its single-valuedness.

We get the next assertion from the meromorphicity of $M_i(z)$ and the decomposition (1.10) (see also [6]).

LEMMA 1.1.1. *Any component $f_j(y)$ of an arbitrary solution $f(y)$ of the system (1.1) with regular singular point a_i can be represented in $\widetilde{O}_i$ as a so-called finite logarithmic sum*

$$f_j(y) = \sum_{k,l} h_{kl}(z)(y - a_i)^{\rho_k} \ln^{b_l}(y - a_i), \tag{1.12}$$

where $0 \le \operatorname{Re}\rho_k < 1$, $b_l \in \mathbb{Z}$, $b_l \ge 0$, the $h_{kl}(z)$ are Laurent series with finite principal part, the sum is finite, and like terms (with respect to the sets of ρ_k and b_l) are combined.

We now proceed to Levelt's [49] description of the structure of the solution space X of the system (1.1) in a neighborhood of a regular singular point (see also [5]).

Suppose that the function $f(y)$ has finite order at a_i.

DEFINITION 1.1.6. The *normalization* $\varphi_i(f)$ of the function $f(y)$ at a_i is defined to be the supremum of the set of integers N such that

$$\operatorname*{Lim}_{y\to a_i}\frac{f(y)}{(y - a_i)^\lambda} = 0 \tag{1.13}$$

for any real number $\lambda < N$. If $f(y) \equiv 0$, then we let $\varphi_i(f) = \infty$ by definition.

It is not hard to show that the number $\varphi_i(f)$ is finite for any function $f(y)$ not identically equal to zero. It follows just as directly from Definition 1.1.6 that for any vector-valued function $f(y)$

$$\varphi_i(f) = \min_{j=1,\dots,p} \varphi_i(f_j), \tag{1.14}$$

where $f_1(y), \dots, f_p(y)$ are the components of this function.

EXAMPLE 1.1.2. Let $h(z)$ be a Laurent series with finite principal part:

$$h(z) = \frac{C_{-m}}{(z-a_i)^m} + \cdots + C_0 + \cdots .$$

Then $\varphi_i(h) = -m$. But if the function $f(y)$ has the form (1.12), then

$$\varphi(f) = \min_{k,l} \varphi_i(h_{kl}). \tag{1.15}$$

For example, the normalizations $\varphi(f_j)$ at 0 of the functions $f_1(z) = z^2 - (1/z)$, $f_2(z) = (1/z^3)y^{1/2}\ln^2 y$, and $f_3(z) = (z, z\ln y, z^2)$ have the respective values

$$\varphi(f_1) = -1, \qquad \varphi(f_2) = -3, \qquad \varphi(f_3) = 1.$$

We consider the solution space X of a system (1.1) with regular singular point a_i on the universal covering $\widetilde{O}_i$. The normalization φ_i gives a mapping $\varphi_i : X|_{\widetilde{O}_i} \to \mathbb{Z} \cup \infty$ having the following properties:

$$\varphi_i(g_i^* f) = \varphi_f(f); \tag{1.16a}$$

$$\varphi_i(f+g) \geq \min(\varphi_i(f), \varphi_i(g)), \tag{1.16b}$$

with equality when $\varphi_i(f) \neq \varphi_i(g)$;

$$\varphi_i(cf) = \varphi_i(f) \quad \text{for any } c \in C \setminus 0. \tag{1.16c}$$

The property (1.16a) follows from the representation of the solution $f(y)$ in the form (1.2), and from (1.15) and (1.8′). Indeed, we consider a component $f_j(y)$ of $f(y)$ for which $\varphi_i(f) = \varphi_i(f_j)$, and in the expansion (1.12) we consider for this component the set σ of pairs (k, l) of indices such that $\varphi_i(h_{kl}) = \varphi_i(f)$ for $(k, l) \in \sigma$. Among the terms of the sum (1.12) with indices in σ we keep those for which $\operatorname{Re}\rho_k$ has the smallest value, and if there are several such terms, then we keep the one for which the number b_l is largest. The term $f_{kl} = h_{kl}(z)(y-a_i)^{\rho_k}\ln^{b_l}(y-a_i)$ singled out is the principal (in the sense of rate of growth) term in the expansion (1.12). From (1.8′),

$$g_i^*(f_{kl}) = \varkappa_\alpha f_{kl} + r(y), \tag{1.11′}$$

where $\varkappa_\alpha$ is an eigenvalue of g_i^*, the powers $\rho_{k'}$ in the sum $r(y)$ are the same, and the powers $b_{l'}$ of the logarithms are less than b_l. Therefore,

$$\varphi_i(g_i^*(f_{kl})) = \varphi_i(f_{kl}) \quad \text{and} \quad \varphi_i(g_i^*(f)) = \varphi_i(f),$$

because the term $\varkappa_\alpha f_{kl}$ in $g_l^*(f_{kl})$ is preserved when combining similar terms.

The property in (1.16b) also follows easily from (1.15) and (1.8′). The property (1.16c) is obvious.

It follows from (1.16b, c) that the normalization φ_i takes a finite number of values $\infty > \varphi_i^1 > \cdots > \varphi_i^l$ on $X|_{\widetilde{O}_i}$ and determines a filtration

$$0 \subset X^1 \subset \cdots \subset X^l = X|_{\widetilde{O}_i} \tag{1.17}$$

of the space $X|_{\widetilde{O}_i}$ by the linear subspaces

$$X^k = \{f \in X|_{\widetilde{O}_i} : \varphi_i(f) \geq \varphi_i^k\}, \qquad k = 1, \dots, l.$$

According to property (1.16a) the operator g_i^* preserves this filtration. Denote by k_m the dimension of the quotient space X^m/X^{m-1}, and by ${}^m g_i^*$ the restriction of g_i^* to X^m.

We consider a basis $e_1^1, \dots, e_{k_1}^1$ of X^1 in which the matrix of the operator ${}^1 g_i^*$ has upper triangular form, complete it to form a basis $e_1^1, \dots, e_{k_1}^1, e_1^2, \dots, e_{k_2}^2$ of X^2 in which the matrix of the operator ${}^2 g_i^*$ has upper triangular form, and so on. (For construction of the elements $e_1^m, \dots, e_{k_m}^m$ it suffices to consider an arbitrary basis $\bar{e}_1^m, \dots, \bar{e}_{k_m}^m$ of the quotient space X^m/X^{m-1} in which the matrix of the operator ${}^{m,m-1}g_i^* : X^m/X^{m-1} \to X^m/X^{m-1}$ induced by ${}^m g_i^*$ has upper triangular form, and then to choose arbitrary representatives of this basis in X^m.)

The basis $(e) = (e_1, \dots, e_p)$ constructed for the space $X|_{\widetilde{O}_i}$ has the following properties:

1) the normalization φ_i takes all its values $\varphi_i^1, \dots, \varphi_i^l$ (counting the multiplicities $k_1, \dots, k_l$) on the elements of the basis (e);

2) $\varphi_i(e_{m+1}) \leq \varphi_i(e_m)$, $m = 1, \dots, p-1$;

3) the matrix G_i of the operator g_i^* has upper triangular form in this basis.

DEFINITION 1.1.7. A basis (e) for the solution space $X|_{\widetilde{O}_i}$ of a system (1.1) with regular singular point a_i satisfying 1)–3) is called an *associated basis*. A basis (e) for $X|_{\widetilde{O}_i}$ satisfying 1) will be said to be *weakly associated*.

It is clear that any associated basis can be obtained in the way described above.

REMARK 1.1.1. Any weakly associated basis is obtained from some associated basis by transformations S_m of the form

$$S_m = \begin{pmatrix} I^{d_1} & 0 & 0 \\ 0 & \widetilde{S}_m & 0 \\ 0 & 0 & I^{d_2} \end{pmatrix}, \tag{1.18}$$

where $d_1 = \dim X^{m-1}$ and $d_2 = p - \dim X^m$ $(m = 1, \dots, l)$, followed by a permutation of elements.

PROOF. We prove an equivalent assertion: from any weakly associated basis (e) we can get an associated basis by successively carrying out permutations and then transformations of the form (1.18). We order the basis (e) in decreasing order of normalization φ_i and get a basis (e') with the properties 1) and 2). Consequently, the first d_m elements of it form a basis of X^m, and the elements $e'_{d_{m-1}+1}, \dots, e'_{d_m}$ are representatives in X^m of a basis $\bar{e}'_{d_{m-1}+1}, \dots, \bar{e}'_{d_m}$ for X^m/X^{m-1}. In each X^m/X^{m-1} we consider a transformation $\widetilde{S}_m$ reducing the operator ${}^{m,m-1}g_i^*$ to upper triangular form, and we form the matrices S_m in (1.18) with the given $\widetilde{S}_m$. It is clear that the basis (e'') obtained from (e') by the transformation $\prod_{m=l}^1 S_m$ has the properties 1)–3), that is, is associated.

EXAMPLE 1.1.3. Suppose that the operator g_i^* reduces to a Jordan block. Then the corresponding Jordan basis (e) will be an associated basis of the space $X|_{\widetilde{O}_i}$.

PROOF. Denote by Y^m the m-dimensional subspace of $X|_{\widetilde{O}_i}$ spanned by the vectors $e_1, \dots, e_m$ of the Jordan basis (e). The filtration

$$0 \subset Y^1 \subset \cdots \subset Y^p = X|_{\widetilde{O}_i} \tag{1.19}$$

is the unique filtration of $X|_{\widetilde{O}_i}$ of length p invariant under the action of g_i^*, since the elements $e_1, \dots, e_p$ form a chain of associated vectors, that is,

$$g_i^*(e_m) = \lambda e_m + e_{m-1}, \qquad m = 2, \dots, p, \ \lambda \in C.$$

In other words, any d_m-dimensional subspace X^m of $X|_{\widetilde{O}_i}$ invariant under the action of g_i^* coincides with the Y^m in (1.19). From this it follows immediately that the Jordan basis (e) is associated with the filtration (1.17), that is, has the properties 1) and 2). Since (e) also has the property 3), it follows that (e) is an associated basis of X.

REMARK 1.1.2. If the matrix of the operator g_i^* reduces to a normal Jordan form containing at most two Jordan blocks with the same eigenvalue λ, then the corresponding Jordan basis (e) is not necessarily weakly associated. As an example we consider the matrix

$$T(y) = \begin{pmatrix} 2\pi i z^3 & z^3 \ln y & \frac{1}{4\pi i} z^3 \ln^2 y - \frac{1}{2} z^3 \ln y & 0 \\ 0 & 2\pi i z^2 & z^2 \ln y & 0 \\ 0 & 0 & 1 & 0 \\ 0 & 2\pi i z & z \ln y & -2\pi i z \end{pmatrix} \tag{1.20}$$

in a neighborhood of the point $a_1 = 0$. In the basis $t_1, \dots, t_4$ of columns of $T(y)$ the monodromy matrix of g_1^* has the form

$$G_1 = \begin{pmatrix} 1 & 1 & 0 & 0 \\ 0 & 1 & 1 & 0 \\ 0 & 0 & 1 & 0 \\ 0 & 0 & 0 & 1 \end{pmatrix},$$

and the normalizations of the columns are equal to

$$\varphi_1(t_1) = 3, \qquad \varphi_1(t_2) = 1, \qquad \varphi_1(t_3) = 0, \qquad \varphi_1(t_4) = 1.$$

In this case the Jordan basis (t) is not weakly associated, because $\varphi_1(t_2 + t_4) = 2 \neq \varphi_1(t_i)$. As an associated basis we can consider the basis t_1, $t_2 + t_4$, t_4, t_3, in which the monodromy matrix has the form

$$\widetilde{G}_1 = \begin{pmatrix} 1 & 1 & 0 & 0 \\ 0 & 1 & 0 & 1 \\ 0 & 0 & 1 & 1 \\ 0 & 0 & 0 & 1 \end{pmatrix}.$$

Any other associated basis is obtained from this by an upper triangular substitution. It follows easily from Remark 1.1.1 that no Jordan basis of the space $\mathscr{L}(t_1, \dots, t_4)$ is weakly associated.

The solution space X on the covering $\widetilde{O}_i$ can be represented as the direct sum of the p_m-dimensional root subspaces ${}^mX$ corresponding to the distinct eigenvalues

$\varkappa_m$ of the operator g_i^*. For each ${}^mX$ we consider the filtration analogous to (1.17), along with some associated basis $({}^me)$ of this space.

DEFINITION 1.1.8. A basis (e) obtained by combining the bases $({}^me)$ over all m will be called a *Levelt basis* of the space X.

LEMMA 1.1.2. *A Levelt basis of X is weakly associated. Any associated basis of X can be obtained by a reshuffling of some associated bases $({}^me)$ of ${}^mX$ and a subsequent upper triangular transformation.*

PROOF. We start with the first assertion of the lemma. Assume the opposite. Suppose that the normalization φ_i either does not take some value φ_i^l on the elements of the Levelt basis (e), or takes it with multiplicity $r < k_l$, where k_l is the multiplicity with which this value is taken by the normalization on the whole of $X|_{\widetilde{O}_i}$ (that is, $k_l = \dim(X^l/X^{l-1})$ in (1.17)). In either case there is an element $f \in X|_{\widetilde{O}_i}$ such that

$$f = f_{i_1} + \cdots + f_{i_m}, \qquad f_{i_j} \neq 0, \qquad f_{i_j} \in {}^{i_j}X,$$
$$\varphi_i(f_{i_1}) = \cdots = \varphi_i(f_{i_m}) < \varphi_i(f). \tag{1.21}$$

Let s be the minimal number for which

$$(g_i^* - \varkappa_{i_m}\,\mathrm{id})^s f_{i_m} = 0.$$

For an arbitrary element $h \in X$ we denote by h' the image $(g_i^* - \varkappa_{i_m}\,\mathrm{id})^s h$. From the properties a)–c) of a normalization, the inequality $\varkappa_\alpha \neq \varkappa_\beta$ for $\alpha \neq \beta$, and (1.11′) we get

$$\varphi_i[(g_i^* - \varkappa_\alpha\,\mathrm{id})f_\beta] = \varphi_i(f_\beta).$$

Therefore, (1.21) gives us

$$f' = f'_{i_1} + \cdots + f'_{i_{m-1}}, \qquad f'_{i_j} \neq 0, \qquad f'_{i_j} \in {}^{i_j}X,$$
$$\varphi_i(f'_{i_1}) = \cdots = \varphi_i(f'_{i_{m-1}}) < \varphi_i(f').$$

In other words, the number of distinct subspaces ${}^kX$ with elements in the decomposition (1.21) for f has been decreased by 1.

Carrying out the above procedure for the subspace ${}^{i_{m-1}}X$, we again decrease by 1 the number of terms in (1.21), and so on. In finitely many steps we get in place of (1.21) the relations

$$\tilde{f} = \tilde{f}_{i_1}, \qquad \tilde{f}_{i_1} \neq 0, \qquad \tilde{f}_{i_1} \in {}^{i_1}X, \qquad \varphi(\hat{f}_{i_1}) < \varphi(\hat{f}),$$

which is impossible. The first assertion of the lemma is proved.

To prove the second assertion it suffices to show that, given any basis for X in which the matrix G_i of g_i^* has upper triangular form, we can obtain by an upper triangular transformation a basis that is a reshuffling of some bases of the root subspaces ${}^kX$. But this is a standard fact in linear algebra.

2. Structure of the solution space of a system with a regular singular point. Fuchsian systems. A criterion for being Fuchsian

Let us consider the space X of solutions of a system (1.1) with a regular singular point a_i, and an arbitrary weakly associated basis (e) of the space $X|_{\widetilde{O}_i}$. Denote by

G_i the matrix of the operator g_i^* in the basis (e), and by $A_i = \operatorname{diag}(\varphi_i^1, \dots, \varphi_i^p)$ the diagonal integer matrix of normalizations of this basis, that is, $\varphi_i^l = \varphi_i(e_l)$. We consider the matrix E_i in (1.9).

LEMMA 1.2.1. *The matrices* $(z-a_i)^{A_i}G_i(z-a_i)^{-A_i}$ *and* $(z-a_i)^{A_i}E_i(z-a_i)^{-A_i}$ *are holomorphic at* a_i. *The normalization of the function* $(z-a_i)^{A_i}(y-a_i)^{E_i}(z-a_i)^{-A_i}$ *at* a_i *is equal to zero.*

PROOF. We first prove the assertion of the lemma for an associated basis (e). In this case the matrices G_i and E_i have upper triangular form according to Definition 1.1.7, and the inequalities $\varphi_i^k \geq \varphi_i^m$ hold when $k \leq m$ for the normalizations of the basis (e). Since for an arbitrary upper triangular matrix $C = ((c_{km}))$ the element $\tilde{c}_{km}$ of the product $(z-a_i)^{A_i}C(z-a_i)^{-A_i}$ is equal to

$$\begin{aligned} \tilde{c}_{km} &= c_{km}(z-a_i)^{\varphi_i^k - \varphi_i^m}, & k \leq m, \\ \tilde{c}_{km} &= 0, & k > m, \end{aligned} \tag{2.1}$$

the matrix $(z-a_i)^{A_i}C(z-a_i)^{-A_i}$ is thereby holomorphic at a_i; this is true, in particular, for $C = G_i$ and $C = E_i$. The matrix $(y-a_i)^{E_i}$ has upper triangular form, and any nonzero element of it can be written in the form (1.12) with $h_{kl}(z) = \text{const}$. It follows from what was proved above that the matrix $(z-a_i)^{A_i}(y-a_i)^{E_i}(z-a_i)^{-A_i}$ also has upper triangular form, any element of it has the form (1.12), and all the $h_{kl}(z)$ in (1.12) are polynomials in z. For the diagonal elements the corresponding $h_{kl}(z)$ are nonzero constants, and so from Definition 1.1.6 and Example 1.1.2 we get

$$\varphi_i((z-a_i)^{A_i}(y-a_i)^{E_i}(z-a_i)^{-A_i}) = 0.$$

Suppose now that (e') is an arbitrary weakly associated basis. According to Remark 1.1.1, it can be obtained from some associated basis (e) by transformations of the form (1.18) and a subsequent permutation S_σ. Let A_i' be the matrix of normalizations of (e'). It follows from the form of the transformations (1.18) that $A_iS_m = S_mA_i$, because the part of the matrix A_i corresponding to $\widetilde{S}_m$ is the scalar matrix equal to $\varphi_i^m I$. Therefore,

$$A_i' = S^{-1}A_iS, \tag{2.2}$$

where $S = S_\sigma \prod_{m=l}^{1} S_m$ is the transition matrix from the basis (e) to (e'). From this it follows that

$$(z-a_i)^{A_i'}S^{-1}CS(z-a_i)^{-A_i'} = S^{-1}((z-a_i)^{A_i}C(z-a_i)^{-A_i})S$$

is a holomorphic matrix at a_i for $C = G_i$ and $C = E_i$, and that the normalization of the matrix

$$(z-a_i)^{A_i'}(y-a_i)^{S^{-1}E_iS}(z-a_i)^{-A_i'} = S^{-1}((z-a_i)^{A_i}(y-a_i)^{E_i}(z-a_i)^{-A_i})S$$

at a_i is equal to zero.

We introduce the notation

$$L_i = \operatorname*{Lim}_{x\to a_i}(z-a_i)^{A_i}E_i(z-a_i)^{-A_i} + A_i. \tag{2.3}$$

Note that, by (2.1),

$$A_i + E_i = L_i + O(z - a_i). \tag{2.4}$$

If (e) is an associated basis, then L_i is an upper triangular matrix whose diagonal coincides with the diagonal of $A_i + E_i$. Denote the diagonal elements of these matrices by

$$\beta_i^j = \varphi_i^j + \rho_i^j, \qquad j = 1, \dots, p. \tag{2.4$'$}$$

DEFINITION 1.2.1. The numbers β_i^j are called the *exponents of the solution space* X *at the point* a_i, or simply the *ith exponents*.

If (e) is a Levelt basis (see Definition 1.1.8), then the blocks of the matrices $A_i + E_i$ and L_i corresponding to the subspace ${}^mX$ have the following appearance:

$$
{}^m(A_i + E_i) = \left(\begin{array}{ccc|ccc|c}
{}^m\beta_i^1 & & * & & & & \\
 & \ddots & & & * & & \\
 & & {}^m\beta_i^1 & & & & \\
\hline
 & & & {}^m\beta_i^2 & & * & \\
 & 0 & & & \ddots & & \\
 & & & & & {}^m\beta_i^2 & \\
\hline
 & & & & & & \ddots
\end{array}\right),
\tag{2.5}
$$

$$
{}^mL_i = \left(\begin{array}{ccc|ccc|c}
{}^m\beta_i & & * & & & & \\
 & \ddots & & & 0 & & \\
 & & {}^m\beta_i^1 & & & & \\
\hline
 & & & {}^m\beta_i^2 & & * & \\
 & 0 & & & \ddots & & \\
 & & & & & {}^m\beta_i^2 & \\
\hline
 & & & & & & \ddots
\end{array}\right),
$$

Here ${}^m\beta_i^1 = {}^m\rho_i + {}^m\varphi_i^1$, ${}^m\beta_i^2 = {}^m\rho_i + {}^m\varphi_i^2$, etc., and each block of the matrix ${}^mL_i$ corresponds to the quotient space ${}^mX^l/{}^mX^{l-1}$ of the filtration (1.17) for ${}^mX$.

PROPOSITION 1.2.1. *The fundamental matrix* $T_i(y)$ *constructed from a weakly associated basis* (e) *of the solution space* $X|_{\tilde{O}_i}$ *of a system* (1.1) *with regular singular point* a_i *has the decompositon*

$$T_i(y) = U_i(z)(z - a_i)^{A_i}(y - a_i)^{E_i}, \tag{2.6}$$

where the matrix $U_i(z)$ *is single-valued and holomorphic in a neighborhood of* a_i.

PROOF. The single-valuedness of $U_i(z)$ follows from Proposition 1.1.1. To prove holomorphicity it suffices to show that for any $\varepsilon > 0$

$$\operatorname*{Lim}_{y \to a_i} U_i(z)(y - a_i)^{2\varepsilon} = 0. \tag{2.7}$$

But

$$U_i(z)(y-a_i)^{2\varepsilon} = T_i(y)(y-a_i)^{-E_i}(z-a_i)^{-A_i}(y-a_i)^{2\varepsilon} = S_i(y)S_2(y),$$

where

$$S_1(y) = T_i(y)(y-a_i)^{-A_i+\varepsilon I},$$
$$S_2(y) = (z-a_i)^{A_i}(y-a_i)^{-E_i}(z-a_i)^{-A_i}(y-a_i)^{\varepsilon}.$$

Since the jth column of $S_1(y)$ is

$$e_j(y) = (y-a_i)^{-\varphi_i(e_j)+\varepsilon},$$

it follows from Definition 1.1.6 that $\operatorname{Lim}_{y\to a_i} S_1(y) = 0$. From Lemma 1.2.1 we get that $\operatorname{Lim}_{y\to a_i} S_2(y) = 0$; therefore, (2.7) holds.

The proposition is proved.

COROLLARY 1.2.1. *Suppose that some fundamental matrix $T_i(y)$ of the space $X|_{\widetilde{O}_i}$ is represented in the form* (2.6), *where $A_i = \operatorname{diag}(\lambda_1, \dots, \lambda_p)$ is an integer matrix, and $U_i(z)$ and $(z-a_i)^{A_i}E_i(z-a_i)^{-A_i}$ are holomorphic matrices. Then the normalizations $\varphi_i(t_j)$ of the columns $t_1, \dots, t_p$ of $T_i(y)$ satisfy the inequalities $\varphi_i(t_j) \geq \lambda_j$, $j = 1, \dots, p$.*

PROOF. For the proof it suffices to show that for any $\varepsilon > 0$

$$\operatorname*{Lim}_{y\to a_i} T_i(y)(y-a_i)^{-A_i+\varepsilon I} = 0.$$

But

$$T_i(y)(z-a_i)^{-A_i+\varepsilon I} = U_i(z)\widetilde{S}_2(y),$$

where

$$\widetilde{S}_2(y) = (z-a_i)^{A_i}(y-a_i)^{E_i}(z-a_i)^{-A_i}(y-a_i)^{\varepsilon}.$$

Since the matrices $U_i(z)$ and $(z-a_i)^{A_i}E_i(z-a_i)^{-A_i}$ are holomorphic, it follows that $U_i(z)$ is bounded in a neighborhood of a_i, and $\operatorname{Lim}_{y\to a_i} \widetilde{S}_2(y) = 0$, which completes the proof of the corollary.

REMARK 1.2.1. For the associated basis (e) the first column $u_1(z)$ of the matrix $U_i(z)$ in the decomposition (2.6) does not vanish at a_i. Indeed, otherwise, $u_1(z) = v(z)(z-a_i)$, where $v(z)$ is a vector-valued function holomorphic at a_i, and $e_1(y) = v(z)(z-a_i)^{\varphi_i(e_1)+1}(y-a_i)^{\rho_i^1}$ for the first column of $T_i(y)$. But the last equality means that $\varphi_i(e_i) \geq \varphi_i(e_1) + 1$, which is impossible.

We recall the definition of the Fuchsian property for a system (1.1) at the point a_i.

DEFINITION 1.2.2. A system (1.1) is said to be *Fuchsian* at a_i if for the matrix form ω of the system the form $(z-a_i)\omega$ is holomorphic at a_i.

The following criterion for a system with regular singular points to be Fuchsian was obtained by Levelt [49] (for the case of a Levelt basis).

PROPOSITION 1.2.2. *A system* (1.1) *with regular singular point a_i is Fuchsian at a_i if and only if in the decomposition* (2.6) *for an arbitrary weakly associated basis* (e)

$$\det U_i(a_i) \neq 0. \tag{2.8}$$

PROOF. *Sufficiency.* According to (2.6), the matrix form ω of the system (1.1) has the following appearance in a neighborhood of a_i:

$$\omega = dT_i \cdot T_i^{-1} = \left(\frac{dU_i}{dz}U_i^{-1} + \frac{U_i}{z-a_i}(A_i + (z-a_i)^{A_i}E_i(z-a_i)^{-A_i})U_i^{-1}\right) dz. \tag{2.9}$$

We get from (2.8) and Lemma 1.2.1 that $(z-a_i)\omega$ is holomorphic at a_i.

Necessity. Suppose that the form ω for the system (1.1) has in a neighborhood of a_i the appearance

$$\omega = \frac{B^i}{z-a_i} + \psi, \tag{2.10}$$

where ψ is a holomorphic form.

We first consider a Levelt basis (e) and prove the assertion for the matrix $T_i(y)$ constructed with respect to this basis.

From (2.9), (2.10), and (2.3),

$$B^i U_i(a_i) = U_i(a_i)L_i. \tag{2.11}$$

We assume that $\det U_i(a_i) = 0$. Denote by Y the kernel of the transformation $U_i(a_i)$:

$$Y = \{v \in C^p : U_i(a_i)v = 0\}.$$

It follows from (2.11) that the transformation L_i carries Y into itself. Let v be an eigenvector for L_i in Y. Then

$$L_i v = \beta_i^j v \tag{2.12}$$

for some j, where β_i^j is one of the exponents of X at a_i. It follows from (2.12) and the form of the matrix L_i in (2.5) that all the components of the vector v are equal to zero except for the components with indices equal to those of the elements of (e) in ${}^mX^l \setminus {}^mX^{l-1}$ for some m and l. Therefore, for the vector-valued function $w = T_i v \in X|_{\widetilde{O}_i}$ we have that $\varphi_i(w) = \varphi_i^j = {}^m\varphi_i^l$, and for v we have the condition

$$A_i v = \varphi_i^j v, \qquad (z-a_i)^{-A_i}v = v(z-a_i)^{-\varphi_i^j}. \tag{2.13}$$

We denote by R the matrix $\operatorname{diag}(\rho_i^1, \dots, \rho_i^p)$, by $\widetilde{E}_i$ the nilpotent matrix $E_i - R$, and by $\widetilde{A}_i$ the matrix $A_i + R = \operatorname{diag}(\beta_i^1, \dots, \beta_i^p)$. Since (e) is a Levelt basis, the matrices E_i and R commute. Indeed, everything reduces to the commutativity of the blocks of these matrices corresponding to the subspaces ${}^mX$, but on each such ${}^mX$ the matrix ${}^mR = {}^m\rho_i \cdot I$ is scalar. We get from this remark and from (2.3) and (2.4) that

$$\begin{aligned}
(y-a_i)^{\widetilde{A}_i}(y-a_i)^{\widetilde{E}_i}(y-a_i)^{-\widetilde{A}_i} &= (z-a_i)^{A_i}(y-a_i)^{\widetilde{E}_i}(z-a_i)^{-A_i} \\
&= \sum_{k=0}^{p-1} \frac{1}{k!} \ln^k(y-a_i)[(z-a_i)^{A_i}\widetilde{E}_i(z-a_i)^{-A_i}]^k \\
&= I + \sum_{k=1}^{p-1} \frac{1}{k!} \ln^k(y-a_i)[L_i - \widetilde{A}_i + O(z-a_i)]^k \\
&= I + O(z-a_i)\sum_{k=1}^{p-1} \frac{1}{k!} \ln^k(y-a_i) + P(y),
\end{aligned} \tag{2.14}$$

where $P(y)=\sum_{k=1}^{p-1}\frac{1}{k!}\ln^k(y-a_i)(L_i-\widetilde{A}_i)^k$. We remark that in view of (2.12) and (2.13)

$$P(y)v=0. \tag{2.15}$$

It follows from (2.13)–(2.15) that

$$\begin{aligned} w(y)(y-a_i)^{-\beta_i^j} &= T_i(y)v(y-a_i)^{-\beta_i^j} = U_i(z)(y-a_i)^{\widetilde{A}_i}(y-a_i)^{\widetilde{E}_i}(y-a_i)^{-\widetilde{A}_i}v \\ &= U_i(z)v+\left[O(z-a_i)\sum_{k=1}^{p-1}\frac{1}{k!}\ln^k(y-a_i)\right]v. \end{aligned} \tag{2.16}$$

Since $U_i(a_i)v=0$, it follows that $U_i(z)v=O(z-a_i)v$, and we get from (2.16) that for any $\lambda<1$

$$\operatorname*{Lim}_{y\to a_i} w(y)(y-a_i)^{-\beta_i^j-\lambda}=0.$$

But $\beta_i^j=\rho_i^j+\varphi_i^j$; therefore,

$$\operatorname*{Lim}_{y\to a_i} w(y)(y-a_i)^{-\varphi_i^j-(\lambda+\operatorname{Re}\rho_i^j)}=0,$$

and since $\operatorname{Re}\rho_i^j\ge 0$ by assumption, we get finally that for any $\lambda<\varphi_i^j+1$

$$\operatorname*{Lim}_{y\to a_i} w(y)(y-a_i)^{-\lambda}=0,$$

and so $\varphi_i(w)\ge\varphi_i^j+1$, which contradicts the condition $\varphi_i(w)=\varphi_i^j$. Hence, $\det U_i(a_i)\neq 0$.

Suppose now that (e') is an arbitrary associated basis. According to Lemma 1.1.2, it can be obtained from some Levelt basis (e) by a reshuffling S_σ and a subsequent upper triangular transformation S_0. For the matrices A_i' and A_i of normalizations of these bases we have

$$A_i'=S_\sigma^{-1}A_iS_\sigma,\qquad S_0^{-1}(z-a_i)^{A_i'}S_0=F(z)(z-a_i)^{A_i'}, \tag{2.17}$$

where $F(z)$ is a holomorphic upper triangular matrix with ones on the main diagonal. Indeed, since S_0 is an upper triangular matrix and $\varphi_i(e_j')\ge\varphi_i(e_{j+1}')$, $j=1,\dots,p-1$, each component of the kth column t_k of the matrix $S_0^{-1}(z-a_i)^{A_i'}S_0$ is a linear combination of elements like $(z-a_i)^{\varphi_i^l}$, where $\varphi_i^l\ge\varphi_i(e_k)$. But by (2.17), the kth column f_k of $F(z)$ is equal to $t_k(z)(z-a_i)^{-\varphi_i(e_k)}$, and hence $f_k(z)$ is a column holomorphic at a_i.

Let T_i' be the fundamental matrix constructed with respect to the basis (e'). Then, by (2.17),

$$\begin{aligned} T_i' &= T_iS_\sigma S_0 = U_i(z)S_\sigma S_0S_0^{-1}(z-a_i)^{A_i'}S_0(y-a_i)^{E_i'} \\ &= U_i(z)S_\sigma S_0F(z)(z-a_i)^{A_i'}(y-a_i)^{E_i'} = U_i'(z)(z-a_i)^{A_i'}(y-a_i)^{E_i'}, \end{aligned} \tag{2.18}$$

where $\det U_i'(a_i)=\det U_i(a_i)\det S_\sigma\det S_0\det F(a_i)\neq 0$, since $\det F(a_i)=1$.

Suppose now that (e) is an arbitrary weakly associated basis. According to Remark 1.1.1, it can be obtained from some associated basis (e') by a transformation of the form (1.18) and a permutation S_σ. Just as in Lemma 1.2.1, we get that (2.2) holds for the matrices of normalizations of these bases; therefore, for the fundamental matrix $T_i(y)$ constructed with respect to the basis (e) we have from (2.18) that

$$\begin{aligned} T_i = T_i'S = U_i'(z-a_i)^{A_i'}(y-a_i)^{E_i'}S \\ = U_i'S(z-a_i)^{S^{-1}A_i'S}(y-a_i)^{S^{-1}E_i'S} = U_i(z)(z-a_i)^{A_i}(y-a_i)^{E_i}, \end{aligned}$$

where $\det U_i(a_i) = \det U_i'(a_i)\det S \neq 0$.

The proposition is proved.

COROLLARY 1.2.2. *Suppose that the fundamental matrix $T_i(y)$ of the space $X|_{O_i}$ is represented in the form*

$$T(y) = U(z)(z-a_i)^A(y-a_i)^{E_i}, \tag{2.19}$$

where A is a diagonal integer matrix, and $U(z)$ is holomorphically invertible in O_i. If the matrix $(z-a_i)^A E_i (z-a_i)^{-A}$ is holomorphic in O_i, then the system (1.1) *with solution space X is Fuchsian at the point a_i.*

The proof follows easily from (2.9).

REMARK 1.2.2. The fact that some fundamental matrix of the solution space of a Fuchsian system can be represented in a neighborhood of a singular point in the form (2.6), (2.8) with some diagonal integer matrix A_i was first proved by Gantmakher [11]. However, he did not try to interpret the eigenvalues of A_i as rates of growth of solutions at a singular point, nor did he consider the problem of singling out the Fuchsian systems among all systems with regular singular points.

COROLLARY 1.2.3. *The eigenvalues of the matrix B^i in* (2.10) *for a system* (1.1) *Fuchsian at the point a_i coincide with the ith exponents of the solution space X of this system.*

PROOF. Let (e) be a Levelt basis. From (2.10) and (2.11),

$$B^i = U_i(a_i)L_iU_i^{-1}(a_i). \tag{2.20}$$

The corollary now follows from the form (2.5) of the matrix L_i.

PROPOSITION 1.2.3. *The sum Σ of all the exponents of the solution space X of a system* (1.1) *with regular singular points $a_1, \dots, a_n$ is an integer and does not exceed zero*:

$$\Sigma = \sum_{i=1}^{n}\sum_{j=1}^{p}\beta_i^j \leq 0. \tag{2.21}$$

A system (1.1) *with regular singular points is a system of Fuchsian type on CP^1 if and only if*

$$\Sigma = 0. \tag{2.22}$$

PROOF. We consider the form $\operatorname{tr}\omega$ on CP^1. Let T_i be the fundamental matrix of $X|_{\tilde{O}_i}$ constructed from a weakly associated basis (e). Then, by (2.9),

$$\operatorname{tr} w = d\ln\det T_i$$

in a neighborhood of zero. It follows from (2.6) that

$$\det T_i = (y - a_i)^{b_i + \sum_{j=1}^{p} \beta_i^j},$$

where b_i is the order of the zero of $\det U_i(z)$ at a_i. Therefore, $\operatorname{res}_{a_i} \operatorname{tr} \omega = b_i + \sum_{j=1}^{p} \beta_i^j$, $b_i \geq 0$. By the residue theorem, $\sum_{i=1}^{n} b_i + \Sigma = 0$, which implies the first assertion of the proposition and the inequality (2.21).

According to Proposition 1.2.2, the system is Fuchsian at the points $a_1, \dots, a_n$ if and only if $b_1 = \cdots = b_n = 0$. This yields the concluding assertion of the proposition.

3. Preliminary facts. Method of solution. Gauge transformations

The main method used in this monograph is the method of gauge transformations. It consists in passing from the system (1.1) to a system with a new unknown function

$$g = \Gamma(z) f, \tag{3.1}$$

where $\Gamma(z)$ is a meromorphic matrix-valued function on CP^1 that is holomorphically invertible away from the singular points $a_1, \dots, a_n$. Under the substitution (3.1) the system (1.1) passes into the system

$$dg = \omega' g, \qquad \omega' = d\Gamma \cdot \Gamma^{-1} + \Gamma \omega \Gamma^{-1}, \tag{3.2}$$

and the solution space X of (1.1) passes into the space $X' = \Gamma(z) X$ of functions, which is the solution space of the system (3.2). (The space X' is understood to be the space of functions generated by the columns of the matrix $\Gamma(z) T(y)$, where $T(y)$ is a fundamental matrix of the space X.)

The form ω' has singularities at the same points as ω, but the character of these singularities can be different. In some cases the formula (3.2) can be used to trace how the order of the pole of ω at a_i changes in dependence on the form of $\Gamma(z)$ [41], but it is technically complicated to do this (mainly because of the presence of the term $d\Gamma \cdot \Gamma^{-1}$). It is much more effective to follow the change in the structure of the solution space (2.6) under gauge transformations, and this is what we do in what follows.

We proceed to a description of the two types of gauge transformations that will be used in the main.

LEMMA 1.3.1. *For any holomorphic function $v(z)$ not vanishing at a_i and for any polynomial*

$$P\left(\frac{1}{z - a_i}\right) = \frac{c_{-s}}{(z - a_i)^s} + \cdots + c_0 \tag{3.3}$$

there is a polynomial

$$Q\left(\frac{1}{z - a_i}\right) = \frac{d_{-s}}{(z - a_i)^s} + \cdots + d_0 \tag{3.4}$$

such that

$$v(z) Q\left(\frac{1}{z - a_i}\right) = P\left(\frac{1}{z - a_i}\right) + o(1). \tag{3.5}$$

PROOF. We represent $v(z)$ in the form

$$v(z) = e_0 + e_1(z - a_i) + \cdots + e_s(z - a_i)^s + o((z - a_i)^s).$$

The equality (3.5) is equivalent to the system of equations

$$\begin{cases} e_0 d_0 + e_1 d_{-1} + \cdots + e_s d_{-s} = -c_0, \\ \qquad e_0 d_{-1} + \cdots + e_{s-1} d_{-s} = -c_{-1}, \\ \cdots\cdots\cdots\cdots\cdots\cdots\cdots\cdots\cdots\cdots \\ \qquad\qquad\qquad e_0 d_{-s} = -c_{-s}. \end{cases}$$

By assumption $e_0 = v(a_i) \neq 0$, so for any $c_0, \ldots, c_{-s}$ this system has a unique solution in $d_0, \ldots, d_{-s}$.

LEMMA 1.3.2. *Suppose that the lth row $u_l(z)$ of the matrix $U(z)$ has the form*

$$u_l(z) = (z - a_i)^k v_l(z) \tag{3.6}$$

in the neighborhood O_i of the point a_i, where the jth component v_{lj} of the row vector $v_l(z)$ is holomorphic in O_i, and $v_{lj}(a_i) \neq 0$. There exists a meromorphic matrix-valued function $\Gamma(z)$ on CP^1 that is holomorphically invertible away from a_i and is such that the jth column x_j of the matrix $U' = \Gamma U$ has the form

$$x_j(z) = (z - a_i)^k x'_j(z), \tag{3.7}$$

where x'_j is holomorphic in O_i, and $x'_{mj}(a_i) = \delta_{ml}$ for the components x'_{mj} of this column (δ_{ml} is the Kronecker delta).

If the matrix $U(z)$ and the row $v_l(z)$ in (3.6) are holomorphic in O_i, then so are all the elements in U'.

PROOF. We represent an element $u_{mj}(z)$, $m \neq l$, of $U(z)$ in the form

$$u_{mj} = (z - a_j)^k \left(P\left(\frac{1}{z - a_i}\right) + o(1) \right).$$

Applying Lemma 1.3.1 to the function $v(z) = v_{lj(z)}$ and to the polynomial $P = P_m$, we get the polynomial $Q_m(\frac{1}{z-a_i})$. Let us consider the gauge transformation

$$\Gamma(z) = \begin{pmatrix} 1 & 0 & \ldots & 0 & Q_1 & 0 & \ldots & 0 & 0 \\ 0 & & & \vdots & \vdots & \vdots & & & 0 \\ \vdots & & & & Q_{l-1} & & & & \vdots \\ & & & & [v_{lj}(a_i)]^{-1} & & & & \\ & & & & Q_{l+1} & & & & \\ & & & & \vdots & & & & 0 \\ 0 & 0 & \ldots & 0 & Q_p & 0 & \ldots & 0 & 1 \end{pmatrix}. \tag{3.8}$$

It follows from (3.5) that the matrix $U' = \Gamma U$ has the property (3.7). It is also clear that $\Gamma(z)$ is holomorphically invertible away from a_i and meromorphic at that point.

The last assertion of the lemma follows from the fact that for a holomorphic matrix $U(z)$ the degrees of the polynomials Q_m in (3.8) do not exceed k, and thus for any element u_{ls} in (3.6) the function $Q_m u_{ls}$ is holomorphic in O_i.

We consider the space X of solutions of a system (1.1) with regular singular points $a_1, \ldots, a_n$, along with the decomposition (2.6).

The transition from the matrix U_i to $U_i' = \Gamma U_i$ and from the space X to $X' = \Gamma X$, where $\Gamma(z)$ is from (3.8), will be called the *$A(a_i, l, j, k)$ procedure.*[1]

The transition from the space X to $X' = \Gamma X$ by means of the gauge transformation

$$\Gamma(z) = \left(\frac{z-a_j}{z-a_i}\right)^C, \tag{3.9}$$
$$C = \operatorname{diag}(c_{11}, \dots, c_{pp}), \qquad c_{ll} \in \mathbb{Z}, \quad l = 1, \dots, p,$$

will be called the *$B(a_i, a_j, c_{11}, \dots, c_{pp})$ procedure.*

Let us consider a system (1.1) that is Fuchsian at the points $a_1, \dots, \hat{\imath}, \dots, a_n$ with a regular singular point at a_i, along with the decomposition (2.6) for an associated basis (e) of the space $C|_{\tilde{O}_i}$.

LEMMA 1.3.3. *Suppose that the lth row u_l of the matrix U_i in the decomposition (2.6) for the space $X|_{\tilde{O}_i}$ has the form (3.6),*

$$u_l(z) = (z-a_i)^k v_l(z), \tag{3.10}$$

where $k \geq 0$, $v_l(z)$ is a row vector holomorphic in O_i, and $v_{lt}(a_i) \neq 0$. If the inequality $\varphi_i^{t-1} \geq \varphi_i^t + k$ ($\varphi_i^0 \stackrel{\text{def}}{=} \infty$) holds for the normalizations φ_i^j of X, then the space $X' = \Gamma(z)X$ obtained from X by the $A(a_i, l, t, k)$ procedure is the solution space of a system (1.1) that is Fuchsian at the points $a_1, \dots, \hat{\imath}, \dots, a_n$, and the normalizations ${}'\varphi_m^j$ and φ_m^j of the spaces X' and X are connected by the following relations:

$$\begin{aligned} {}'\varphi_m^j &= \varphi_m^j, \qquad j = 1, \dots, p, \ m \neq i, \\ {}'\varphi_i^t &= \varphi_i^t + k, \qquad {}'\varphi_i^j \geq \varphi_i^j, \qquad j = 1, \dots, \hat{t}, \dots, p. \end{aligned} \tag{3.11}$$

PROOF. The Fuchsian property at $a_1, \dots, \hat{\imath}, \dots, a_n$ of a system (1.1) with solution space X' and the first equality in (3.11) follow from Propositions 1.2.1 and 1.2.2 and the fact that the decompositions (2.6) for $X'|_{\tilde{O}_i}$ have the form

$$T_m'(y) = \Gamma(z)U_m(z)(z-a_m)^{A_m}(y-a_m)^{E_m},$$

where $\det\Gamma(a_m) \neq 0$ for $m \neq i$ for the matrix $\Gamma(z)$ in (3.8).

Denote by C the matrix

$$C = \operatorname{diag}(\underbrace{0, \dots, 0}_{t-1}, k, 0, \dots, 0).$$

It follows from Lemma 1.3.1 that the matrix $\Gamma(z)U_i(z)$ can be written as

$$\Gamma(z)U_i(z) = U_i'(z)(z-a_i)^C,$$

where $U_i'(z)$ is holomorphic in O_i, and $u_{lt}'(a_i) \neq 0$. Therefore, the matrix T_i' can be represented in the form

$$T_i'(y) = U_i'(z)(z-a_i)^{A_i+C}(y-a_i)^{E_i}.$$

---

[1]The $A(a_i, l, j, k)$ procedure carries over without any difficulties also to negative values of k.

From the condition $\varphi_i^{t-1} \geq \varphi_i^t + k$ and from Corollary 1.2.1 we get that

$$'\varphi_i^t = \varphi_i^t + k\,, \qquad '\varphi_i^j \geq \varphi_i^j\,, \quad j \neq t.$$

REMARK 1.3.1. If under the condition of Lemma 1.3.3 we omit the condition that the components $v_{l,t+1}(z), \dots, v_{lp}(z)$ of the row vector $v_l(z)$ in (3.10) be holomorphic, then the conditions (3.11) take the form

$$(3.12)\qquad \begin{aligned} &\qquad\qquad\qquad '\varphi_i^j \geq \varphi_i^j\,, \qquad j < t; \\ &'\varphi_i^t = \varphi_i^t + k\,, \qquad '\varphi_i^{t+s} \geq \varphi_i^{t+s} - k_{t+s}\,, \qquad s = 1, \dots, p-t, \end{aligned}$$

where

$$k_{t+1} = \max(0, k - l_{t+1}),\ k_{t+2} = \max(0, k - l_{t+1}, k - l_{t+2}), \dots,$$
$$k_p = \max(0, k - l_{t+1}, \dots, k - l_p),$$

with l_j the order of the zero of $v_{lj}(z)$ at a_i if $v_{lj}(z) \neq 0$, and $l_j = k$ if $v_{lj}(z) = 0$.

The inequalities (3.12) follow from the fact that in this case the matrix $\Gamma(z)U_i(z)$ can be written in the form

$$\Gamma(z)U_i(z) = U_i'(z)(z-a_i)^{C'},$$

where $C' = \operatorname{diag}(0, \dots, 0, k, -k_{t+1}, \dots, -k_p)$, $U_i'(z)$ is a holomorphic matrix on O_i, and $k \geq -k_{t+1} \geq \dots \geq -k_p$.

The following assertion will be used in Chapters II and III.

PROPOSITION 1.3.1. *Suppose that the rows* $u_1, \dots, u_m$ *of an* $m \times m$ *matrix that is holomorphic in the neighborhood* O_l *of* a_l *have the form*

$$u_i(z) = (z - a_l)^{k_i} v_i(z),$$

where $k_i > 0$, *and* $v_i(z)$, $i = 1, \dots, m$, *are holomorphic vector-valued functions on* O_l *such that the vectors* $v_1(a_l), \dots, v_m(a_l)$ *are linearly independent. For any holomorphic vector-valued function* $w(z) = (w^1, \dots, w^m)$ *there exist polynomials* $Q_i(\frac{1}{z-a_i})$ *of degrees* k_i *such that*

$$Q_1\left(\frac{1}{z-a_l}\right)u_1(z) + \dots + Q_m\left(\frac{1}{z-a_l}\right)u_m(z) + w(z) = (z-a_l)^k t(z),$$

where $k \geq \min(k_1, \dots, k_m)$, *and* $t(z)$ *is a holomorphic vector-valued function.*

PROOF. Let $\sum_{i=1}^m c_i^1 v_i(a_l) + w(a_l) = 0$. Then

$$\sum_{i=1}^m c_i^1 \frac{1}{(z-a_l)^{k_i}} u_i(z) + w(z) = z t_1(z),$$

where $t_1(z)$ is a holomorphic vector-valued function. We consider the coordinates c_i^2 of the vector $-t_1(a_l)$ in the basis $(v_i(a_l))$. Then

$$\sum_{i=1}^m c_i^2 \frac{1}{(z-a_l)^{k_i-1}} u_i(z) + z t_1(z) = z^2 t_2(z),$$

where $t_2(z)$ is a holomorphic vector-valued function, and so on. The polynomial $Q_i(\frac{1}{z-a_l})$ has the form

$$Q_i\left(\frac{1}{z-a_l}\right) = c_i^1 \frac{1}{(z-a_l)^{k_i}} + c_i^2 \frac{1}{(z-a_l)^{k_i-1}} + \cdots.$$

In what follows we need some assertions about the system (1.1) that have to do with reducibility of its monodromy (1.5).

LEMMA 1.3.4. *If some component $f_j(y)$ of a function $f(y)$ in the solution space X of a system* (1.1) *with monodromy* (1.5) *satisfies the identity $f_j(y) \equiv 0$, then the space* (1.5) *is reducible.*

PROOF. Consider the l-dimensional subspace $X_l \subset X$ consisting of the functions $f(y)$ with jth components identically equal to zero. By an assumption of the lemma, $0 < l < p$. According to (1.3), for any monodromy operator g^* and any $f \in X_l$

$$(g^* f)_j(y) = f_j(g^{-1} y) \equiv 0.$$

This means that X_l is a common invariant subspace for the monodromy operators, and hence the monodromy representation (1.5) of (1.1) is reducible.

LEMMA 1.3.5. *If the condition*

$$\omega_{ij} \equiv 0, \qquad i = l+1, \dots, p, \; j = 1, \dots, l, \; l < p, \tag{3.13}$$

holds for the matrix differential form ω of the system (1.1), *then the monodromy representation* (1.5) *of* (1.1) *is reducible.*

PROOF. Consider the system $df = \omega' f$, where $\omega' = \|\omega_{ij}\|$, $1 \le i, j \le l$. If f' is a solution of this system, then the vector-valued function $f = (f', 0, \dots, 0)$ is a solution of the original system (1.1). Lemma 1.3.5 now follows from Lemma 1.3.4.

We again consider a system (1.1) that is Fuchsian at the points $a_1, \dots, \hat{\imath}, \dots, a_n$ and has a regular singular point at a_i, and we consider the decomposition (2.6) for the solution space X of this system.

LEMMA 1.3.6. *Suppose that the matrix $U_i(z)$ in the decomposition* (2.6) *for the space $X|_{\tilde{O}_i}$ has the form*

$$U_i(z) = (z-a_i)^C V(z), \tag{3.14}$$

where $C = \operatorname{diag}(0, \dots, 0, c_{l+1,l+1}, \dots, c_{pp})$, $c_{mm} \in \mathbb{Z}$ and $c_{mm} \ge 0$ for all m, and the matrix $V(z) = \|v_{km}\|$ is holomorphically invertible in the neighborhood O_i of a_i. If the elements u^j_{km} of the matrices U_j $(j = 1, \dots, n)$ in the decomposition (2.6) *at the points $a_1, \dots, a_n$ and the elements v_{km} of the matrix V satisfy the equalities*

$$\begin{gathered} u^j_{km}(a_j) = 0, \qquad v_{km}(a_i) = 0, \\ l+1 \le k \le p, \qquad 1 \le m \le l, \qquad j = 1, \dots, n, \end{gathered} \tag{3.15}$$

then the monodromy representation (1.5) *of the system* (1.1) *is reducible.*

PROOF. It follows from the form of the matrix $U_i(z)$ that the elements u^i_{km} of $U_i^{-1}(z)$ with $1 \le k \le p$ and $1 \le m \le l$ are holomorphic in the neighborhood O_i of

a_i, and (3.15) holds for them. It is clear that (3.15) holds also for the elements of the matrices $U_1^{-1}(z), \dots, \hat{\imath}, \dots, U_n^{-1}(z)$, which are holomorphic in a neighborhood of the respective points $a_1, \dots, \hat{\imath}, \dots, a_n$ in view of Proposition 1.2.2. We get from (2.9) that the elements ω_{km} of the matrix differential form ω are holomorphic at all the points $a_1, \dots, a_n$ for $l+1 \le k \le p$ and $1 \le m \le l$. Since ω is holomorphic on $CP^1 \setminus \mathscr{D}$, this means that the forms ω_{km} are holomorphic on the whole of CP^1, and therefore $\omega_{km} \equiv 0$ for the indicated indices k and m. Lemma 1.3.6 now follows from Lemma 1.3.5.

LEMMA 1.3.7. *Suppose that the monodromy representation* (1.5) *of a system* (1.1) *with regular singular points* $a_1, \dots, a_n$ *is reducible and that the subspace* X_l *of the solution space* X *of this system is a common invariant subspace for the monodromy operators. Then the sum* Σ_l *of the exponents of* X_l *with respect to all the points* $a_1, \dots, a_n$ *is an integer and satisfies the inequality*

$$\Sigma_l = \sum_{i=1}^{n} \sum_{j=1}^{l} \beta_i^j \le 0. \tag{3.16}$$

PROOF. Choose a basis $(e_1, \dots, e_l)$ in X_l and consider the fundamental matrix $T(y)$ constructed with respect to this basis. Denote by $T'(y)$ the matrix formed from the elements of a basic minor of $T(y)$ at a point $y_0 \in \widetilde{CP^1 \setminus \mathscr{D}}$. Then $\det T'(y_0) \ne 0$. The space X' generated by the columns of $T'(y)$ is the solution space of the system $df = \omega' f$ of l linear differential equations, where $\omega' = dT'(T')^{-1}$. The set of singular points of the system consists of the points $a_1, \dots, a_n$ and additional "false" singularities $a'_{n+1}, \dots, a'_r$. The latter includes the points for which $\det T'(y) = 0$, $y \in \pi^{-1}(a'_t)$. We remark that if $\det T'(y) = 0$ for some $y \in \pi^{-1}(a'_t)$, then $\det T'(y) = 0$ for any $y \in \pi^{-1}(a'_t)$ in view of (1.3).

It follows from the last remark that the number of additional singular points is finite, for otherwise either there is a limit point $\tilde{y} \in \widetilde{CP^1 \setminus \mathscr{D}}$ for the set $\{\pi^{-1}(a'_{n+i})\}$, or the set $\{a'_{n+i}\}$ has one of the singular points a_j $(j = 1, \dots, n)$ as its limit point. In the first case the uniqueness theorem for analytic functions gives us that $\det T'(y) \equiv 0$ when applied to $\det T'(y)$, and in the second case the same theorem gives us the same result when applied to the function $\det T'(y)(y - a_i)^{-\operatorname{tr} E'_j}$, where $\exp(2\pi i E'_j)$ is the monodromy of $T'(y)$ (note that this function is meromorphic at a_j); this contradicts the condition $\det T'(y_0) \ne 0$.

The exponents ${}'\beta^j_{n+i}$ of X' at the points a'_{n+i} coincide with the normalizations ${}'\varphi^j_{n+i}$, which in turn are nonnegative because $T'(y)$ is analytic at the points $\pi^{-1}(a'_{n+i})$:

$$'\beta^j_{n+i} = {}'\varphi^j_{n+i} \ge 0, \qquad 1 \le i \le r. \tag{3.17}$$

The normalizations ${}'\varphi^j_i$ of X' are connected with the normalizations φ^j_i of X_l at the points $a_1, \dots, a_n$ by the inequalities

$$'\varphi^j_i = \varphi_i(e'_j) \ge \varphi_i(e_j) = \varphi^j_i, \tag{3.18}$$

which follow from the definition of normalizations, the fact that the column vector $e'_j(y)$ of the matrix $T'(y)$ is obtained from the vector-valued function $e_j(y)$ by omitting a certain number of components, and from (1.14).

From Proposition 1.2.3 and the inequalities (3.17) and (3.18) we get that

$$\Sigma_l = \sum_{i=1}^{n}\sum_{j=1}^{l}\beta_i^j \le \sum_{i=1}^{n}\sum_{j=1}^{l}{}'\beta_i^j + \sum_{i=n+1}^{n+r}\sum_{j=1}^{l}{}'\varphi_i^j = \Sigma'_l \le 0,$$

where Σ'_l is the sum of the exponents of the space X'.

LEMMA 1.3.8. *Suppose that a representation* (1.5) *of dimension* $p = 3$ *is reducible, and all the monodromy operators have a common invariant subspace of dimension* $l = 1$ *or* $l = 2$. *Then there exist a system* (1.1) *with regular singular points* $a_1, \dots, a_n$ *that is Fuchsian at* $a_2, \dots, a_n$ *and whose monodromy coincides with* (1.5), *and a fundamental matrix* $T(y)$ *of the solution space* X *of this system such that the elements* e_{km} *of* $T(y)$ *satisfy the identities*

$$e_{km}(y) \equiv 0, \qquad l+1 \le k \le 3, \ 1 \le m \le l. \tag{3.19}$$

PROOF. Consider a system (1.1) with regular singular points $a_1, \dots, a_n$ that is Fuchsian at $a_2, \dots, a_n$ and whose monodromy coincides with (1.5). The existence of such a system was proved in [52]; see also Proposition 1.4.1. In the space $X|_{\tilde{O}_1}$ it is always possible to choose a weakly associated basis (e) in which the matrix G_1 of the monodromy operator g_1^* has upper triangular form and for which the first l elements are an associated basis for $X_l|_{\tilde{O}_1}$. Indeed, the filtration (1.17) induces a filtration in $X_l|_{\tilde{O}_1}$, and the operator $g_1^*|_{X_1}$ preserves this filtration. We consider an associated basis $(e_1, \dots, e_l)$ of $X_l|_{\tilde{O}_1}$ and complete it to form a basis in $X|_{\tilde{O}_1}$ as follows. The elements $(e_1, \dots, e_l)$ are supplemented by elements $e_1^1, \dots, e_{s_1}^1$ whose images $\bar{e}_1^1, \dots, \bar{e}_{s_1}^1$ in the quotient space $X^1/X^1 \cap X_l$ form a basis in which the induced operator g_1^* has upper triangular form (here X^1 is a subspace in the filtration (1.17)). The same is done for $X^2/X^2 \cap X_l$, and so on. As a result we get a basis (e).

It can be assumed that the equalities

$$u_{21}(a_1) = u_{31}(a_1) = u_{32}(a_1) = 0 \tag{3.20}$$

hold for the elements u_{km} of $U_1(z)$ in the decomposition (2.6) for the fundamental matrix $T(y)$ constructed from the basis (e). The conditions (3.20) can always be satisfied with the help of linear transformations of the rows of $U_1(z)$; this corresponds to passing from a system (1.1) with solution space X to a system with solution space $X' = SX$, where S is a nonsingular constant matrix.

1. Let us consider first the case $l = 1$. Assume that

$$u_{j1}(z) \not\equiv 0, \qquad j = 2 \text{ or } j = 3. \tag{3.21}$$

Denote by r the order of the zero of $u_{j1}(z)$ at a_1; it follows from (3.20) that $r > 0$.

We apply the $A(a_1, j, 1, r)$ procedure to X and get a space $X' = \Gamma_1(z)X$. According to Remark 1.3.1, the sum Σ' of the exponents of the space $X'_1 = \Gamma_1(z)X_1$ is connected with the sum Σ of the exponents of X_1 by the inequality

$$\Sigma' = \Sigma + r > \Sigma. \tag{3.22}$$

We again reduce the space X' to the form (3.20). If (3.21) holds for some element $u'_{j1}(z)$ of $U'_1(z)$, then again we use the $A(a_1, j, 1, r')$ procedure, and so on. It follows from Lemma 1.3.7 and from (3.22) that in finitely many steps we obtain the solution space $\widetilde{X}$ of a system (1.1) with regular singular points $a_1, \dots, a_n$ that is Fuchsian at $a_2, \dots, a_n$ and has the given monodromy (1.5), and $\widetilde{X}$ is such that the identities $u_{21}(z) \equiv u_{31}(z) \equiv 0$ hold for the elements of the matrix $\widetilde{U}_1(z)$ in the decomposition (2.6), and hence also $e_{21} \equiv e_{31}(z) \equiv 0$ (since E_1 is upper triangular).

2. Let us now consider the case $l = 2$. Two subcases are possible:

a) the two-dimensional monodromy representation χ^2 of the space X_2 is reducible;

b) χ^2 is irreducible.

According to the analyzed case $l = 1$ it can be assumed in case a) that $T(y)$ already has the form

$$T(y) = \begin{pmatrix} e_{11} & * & * \\ 0 & \multicolumn{2}{c}{\widetilde{T}_2(y)} \\ 0 & & \end{pmatrix}, \tag{3.23}$$

where $\widetilde{T}_2(y)$ is the fundamental matrix of the solution space of a system (1.1) whose monodromy representation coincides with the corresponding factor representation of (1.5). This factor representation has a one-dimensional invariant subspace generated by the vector $e_2 = (e_{22}, e_{32})^T$; therefore, the analysis of the case $l = 1$ implies the existence of a matrix $\Gamma'(z)$ such that the equality $t_{21}(z) \equiv 0$ holds for the element t_{21} of the matrix $\widetilde{T}'_2 = \Gamma' \widetilde{T}_2$. We consider the matrix

$$\Gamma(z) = \begin{pmatrix} 1 & 0 & 0 \\ 0 & \multicolumn{2}{c}{\Gamma'(z)} \\ 0 & & \end{pmatrix}.$$

For the matrix $T'(y) = \Gamma(z) T(y)$ we get that

$$e'_{21}(y) \equiv e'_{31}(y) \equiv e'_{32}(y) \equiv 0. \tag{3.24}$$

In case b) we assume that $u_{31}(z) \not\equiv 0$, and we denote by r_1 and r_2 the orders of the zeros of the respective functions $u_{31}(z)$ and $u_{32}(z)$ at a_1. It follows from (3.20) that $r_1 > 0$ and $r_2 > 0$ (if $u_{32}(z) \equiv 0$, then we set $r_2 = r_1$). Applying the $A(a_1, 3, 1, r_1)$ procedure to X, we get the space $X' = \Gamma_1(z) X$. According to Remark 1.3.1, the sum Σ' of the exponents of $X'_2 = \Gamma(z) X_2$ is connected with the sum Σ of the exponents of X_2 by the inequality

$$\Sigma' \geq \Sigma + r_1 - \max(0, r_1 - r_2) > \Sigma. \tag{3.25}$$

We again reduce the space X' to the form (3.20). If the element $u'_{31}(z)$ of the matrix $U'_1(z) = \Gamma_1 U_1(z)$ is not identically equal to zero, then we again employ the $A(a_1, 3, 1, r'_1)$ procedure, and so on. It follows from Lemma 1.3.7 and (3.25) that in finitely many steps we get the solution space $\widetilde{X}$ of a system (1.1) with regular singular points $a_1, \dots, a_n$ that is Fuchsian at $a_2, \dots, a_n$ with the given monodromy (1.5), and $\widetilde{X}$ is such that the identity $\widetilde{u}_{31}(z) \equiv 0$ holds for the element $\widetilde{u}_{31}(z)$ of the matrix $\widetilde{U}_1(z)$ in the decomposition (2.6). Therefore, $e_{31}(y) \equiv 0$, and $\widetilde{e}_{32}(y) \equiv 0$ because χ^2 is irreducible (otherwise the vector $e_1(y)$ is a common eigenvector for the monodromy operators in view of Lemma 1.3.4, and χ^2 is reducible).

COROLLARY 1.3.1. *If all the monodromy matrices of a p-dimensional representation* (1.5) *can be reduced simultaneously to upper triangular form, then there exists a system* (1.1) *with regular singular points $a_1, \dots, a_n$ that is Fuchsian at $a_2, \dots, a_n$ and has monodromy coinciding with* (1.5), *and the fundamental matrix $T(y)$ of the solution space of this system has upper triangular form.*

The proof is easy to carry out by induction; the first step was reproduced in the analysis of case 2(a) in Lemma 1.3.8.

3. The following lemma will see constant use in what follows.

LEMMA 1.3.9. *For any matrix $U(z)$ holomorphic in the neighborhood O_i of a_i and holomorphically invertible in $O_i \setminus a_i$ there exists a matrix-valued function $\Gamma(z)$ that is meromorphic on CP^1 and holomorphically invertible away from a_i such that*

$$\Gamma(z)U(z) = (z-a_i)^C V(z), \tag{3.26}$$

where $\det \Gamma(z) \equiv \text{const}$, *$V(z)$ is a matrix that is holomorphically invertible in O_i, $C = \operatorname{diag}(c_{11}, \dots, c_{pp})$, $c_{ll} \in \mathbb{Z}$, and $0 \le c_{11} \le \dots \le c_{pp}$.*

PROOF. Denote by k the order of the zero of $\det U(z)$ at a_i, and suppose that $k > 0$.

Assume that the matrix $U(z)$ is already reduced to the form

$$U_1(z) = \Gamma_1(z)U(z) = (z-a_i)^{C_1} V_1(z), \tag{3.27}$$

where $\Gamma_1(z)$ and C_1 satisfy the conditions of the lemma, $V_1(z)$ is holomorphic in O_i and holomorphically invertible away from a_i, and the function $\det V_1(z)$ has a zero of order $k_1 \le k$ at a_i. Then

$$0 \le \operatorname{tr} C_1 \le k.$$

(Note that the original matrix $U(z)$ has the form (3.27) with $\Gamma_1 = I$ and $C_1 = 0 \cdot I$.)

To prove the lemma it suffices to show that $U_1(z)$ can be reduced to the form (3.27) by a matrix $\Gamma_2(z)$ satisfying the conditions of the lemma, with the matrices U_1 and V_1 replaced by $\widetilde{U}$ and $\widetilde{V}$ and the matrix C_1 replaced by a matrix $\widetilde{C}_1$ such that $\operatorname{tr} \widetilde{C}_1 > \operatorname{tr} C_1$.

The row ${}^1u_l(z)$ of the matrix $U_1(z)$ has the form

$${}^1u_l(z) = (z-a_i)^{{}^1c_{ll}} \cdot {}^1v_l(z),$$

where ${}^1v_l(z)$ is a row of the matrix $V_1(z)$. Suppose that the element ${}^1v_{pj_p}(a_i) = b_p$ is nonzero. Let

$$\Gamma_2(z) = \begin{pmatrix} 1 & 0 & \dots & 0 & -(z-a_i)^{{}^1c_{11} - {}^1c_{pp}} \cdot \frac{v_{1jp}(a_i)}{b_p} \\ 0 & 1 & \cdots & 0 & \\ \vdots & & & \vdots & \vdots \\ 0 & 0 & \dots & 1 & -(z-a_i)^{{}^1c_{p-1\,p-1} - {}^1c_{pp}} \cdot \frac{v_{p-1,j_p}(a_i)}{b_p} \\ 0 & 0 & \dots & 0 & 1 \end{pmatrix}.$$

The matrix $\Gamma_2(z)$ is holomorphically invertible away from the point a_i, $\det \Gamma_2 = 1$, and the matrix $U_2(z) = \Gamma_2(z)U_1(z)$ can be represented in the form

$$U_2(z) = (z-a_i)^{C_1} V_2(z),$$

where the j_pth column ${}^2x_{j_p}$ of the holomorphic matrix $V_2(z)$ has at a_i the form

$$ {}^2x_{j_p}(a_i) = (0, \dots, 0, b_p)^T . $$

For the $(p-1)$st row ${}^2v_{p-1}$ of $V_2(z)$ there are two possibilities:

a) ${}^2v_{p-1}(a_i) = 0$;

b) ${}^2v_{p-1, j_{p-1}}(a_i) = b_{p-1} \neq 0$ for some $j_{p-1} \neq j_p$.

In case a)

$$ V_2(z) = (z - a_i)^{C_2} \widehat{V}(z), $$

where $C_2 = \operatorname{diag}(0, \dots, 0, t, 0)$, the matrix $\widehat{V}(z)$ is holomorphic at zero, and t is the order of the zero of the vector-valued function ${}^2v_{p-1}(z)$ at a_i. Therefore, the matrix $U_2(z)$ can be written in the form

$$ U_2(z) = (z - a_i)^{\widehat{C}} \widehat{V}(z), $$

where $\operatorname{tr} \widehat{C} = \operatorname{tr} C_1 + t > \operatorname{tr} C$, and this completes the proof.

In case b) we consider the matrix

$$ \Gamma_3(z) = \begin{pmatrix} 1 & 0 & \dots & 0 & -(z-a_i)^{{}^1c_{11} - {}^1c_{p-1\,p-1} \cdot \frac{v_{1 j_{p-1}}(a_i)}{b_{p-1}}} & 0 \\ 0 & 1 & & & & \\ \vdots & & \ddots & \vdots & \vdots & \vdots \\ 0 & & \cdots & 1 & -(z-a_1)^{{}^1c_{p-2\,p-2} - {}^1c_{p-1\,p-1} \cdot \frac{{}^2v_{p-2 j_{p-1}}(a_i)}{b_{p-1}}} & \\ 0 & & & & 1 & 0 \\ 0 & & & & 0 & 1 \end{pmatrix} $$

The matrix $U_3(z) = \Gamma_3(z) U_2(z)$ can be represented as

$$ U_3(z) = (z - a_i)^{C_1} V_3(z), $$

where the j_{p-1}th column ${}^3x_{j_{p-1}}$ and the j_pth column ${}^3x_{j_p}$ of the matrix $V_3(z)$ have at a_i the forms

$$ {}^3x_{j_p}(a_i) = (0, \dots, 0, b_p)^T, $$
$$ {}^3x_{j_{p-1}}(a_i) = (0, \dots, 0, b_{p-1}, *)^T . $$

For the $(p-2)$nd row ${}^3v_{p-2}$ of $V_3(z)$ there are again the possibilities a) and b), and so on. Either a) holds at some step for some row, or, carrying out the above procedure for all rows, we get as a result the matrix

$$ \widetilde{U} = (z - a_i)^{C_1} \widetilde{V}(z), $$

where the column $\tilde{x}_{j_l}$ of the holomorphic matrix $\widetilde{V}(z)$ has the form

$$ \tilde{x}_{j_l}(a_i) = (0, \dots, 0, b_l, *, \dots, *)^T, \qquad b_l \neq 0. $$

But in this case $\det \widetilde{V}(a_i) = \prod_{l=1}^{p} b_l \neq 0$ and $\operatorname{tr} C_1 = k$, and hence the matrix $U_1(z)$ in (3.27) already has the required form. The lemma is proved.

LEMMA 1.3.10. *Under the conditions of Lemma* 1.3.9 *the matrix* C *in the decomposition* (3.26) *is uniquely determined by the matrix* $U(z)$.

PROOF. Assume the contrary. Suppose that two decompositions (3.26) with different C_1 and C_2 hold for the matrix $U(z)$:

$$\Gamma_1(z)U(z) = (z-a_i)^{C_1}V_1(z),$$
$$\Gamma_2(z)U(z) = (z-a_i)^{C_2}V_2(z).$$

Denote by l the first index with ${}^1c_{ll} \neq {}^2c_{ll}$. For definiteness assume that

$$ {}^1c_{ll} > {}^2c_{ll}. \tag{3.28}$$

Then

$$\Gamma_1 \cdot \Gamma_2^{-1} = (z-a_i)^{C_1}D(z)(z-a_i)^{-C_2}, \tag{3.29}$$

where $D(z) = V_1(z)V_2^{-1}(z)$ is holomorphically invertible in O_i. Since the matrix $\Gamma_1\Gamma_2^{-1}$ is holomorphically invertible away from a_i, the matrix $D(z)$ is meromorphic on CP^1 and holomorphically invertible away from the point ∞. We consider the element $\gamma_{km}(z)$ of $\Gamma_1(z)\Gamma_2^{-1}(z)$ with indices $1 \leq k \leq l$, $l \leq m \leq p$. By assumption, $\gamma_{km}(z)$ is holomorphic away from a_i. But

$$\gamma_{km}(z) = (z-a_i)^d d_{km}(z),$$

where $d = {}^1c_{kk} - {}^2c_{mm} > 0$ according to (3.28), and $d_{km}(z)$ is an element of $D(z)$. Therefore, the function $\gamma_{km}(z)$ is holomorphic on the whole Riemann sphere, and $\gamma_{km}(a_i) = 0$. Hence, $\gamma_{km}(z) \equiv 0$ for $1 \leq k \leq l$ and $l \leq m \leq p$. But then $\det \Gamma_1(z)\Gamma_2^{-1}(z) \equiv 0$, which contradicts the holomorphic invertibility of this matrix away from a_i,

$$\Gamma_1(z)\Gamma_2^{-1}(z) = \left(\begin{array}{cc|ccc} \ddots & * & 0 & \cdots & 0 \\ * & \ddots & \vdots & 0 & \vdots \\ \hline & & 0 & \cdots & 0 \\ & * & & \ddots & * \\ & & & & \ddots \end{array}\right) - l.$$

The lemma is proved.

REMARK 1.3.2. Lemma 1.3.9 is a modification of a lemma of Sauvage (see [26] or [2]) and can be used together with Lemma 1.3.10 for an elementary proof of the Birkhoff–Grothendieck theorem (4.14).

4. The Riemann–Hilbert problem and holomorphic bundles on the Riemann sphere

In this section we use the notation from the first part of §1.

The universal covering $R = (\widetilde{CP^1 \setminus \mathscr{D}}, CP^1 \setminus \mathscr{D}, \pi)$ can be regarded as a principal Δ-bundle, with the covering transformation group Δ acting on $\widetilde{CP^1 \setminus \mathscr{D}}$ from the left (see [27]). We define a right action of Δ as follows:

$$y\sigma = \sigma^{-1}y, \qquad y \in \widetilde{CP^1 \setminus \mathscr{D}}, \ \sigma \in \Delta. \tag{4.1}$$

Denote by $\widetilde{F} = (\widetilde{F}_E, CP^1\backslash\mathscr{D}, \widetilde{\pi}_F)$ the vector bundle with structure group $GL(p; C)$ associated with R by (1.5). (Recall that the fundamental group $\pi_1(CP^1 \backslash \mathscr{D}, z_0)$ is identified with Δ.) The bundle $\widetilde{F}$ admits a simple coordinate description with constant transition functions. Such a description is given in [53] and [25], and in the same place it is proved that $\widetilde{F}$ is holomorphically trivial. Therefore, so is the principal $GL(p; C)$-bundle $\widetilde{P} = (\widetilde{P}_E, CP^1 \backslash \mathscr{D}, \widetilde{\pi}_p)$, with which $\widetilde{F}$ is also associated. The total space $\widetilde{P}_E$ of this bundle can be described as follows:

$$\widetilde{P}_E = (\widetilde{CP^1\backslash\mathscr{D}}) \times GL(p; C)/\sim, \tag{4.2}$$

where

$$(y; G) \sim (y\sigma; \chi^{-1}(\sigma)G) \sim (\sigma^{-1}y; \chi^{-1}(\sigma)G) \sim (\sigma y; \chi(\sigma)G),$$
$$y \in \widetilde{CP^1\backslash\mathscr{D}}, \quad G \in GL(p; C), \quad \sigma \in \Delta.$$

Denote by $\langle y; G\rangle$ the equivalence class of an element $(y; G)$ with respect to $\sim$.

We consider the mapping

$$W: \widetilde{CP^1\backslash\mathscr{D}} \to \widetilde{P}_E, \qquad W(y) = \langle y; I\rangle. \tag{4.3}$$

Let $U(z)\colon CP^1 \backslash \mathscr{D} \to \widetilde{P}_E$ be a holomorphic section of the bundle $\widetilde{P}$. Then

$$W(y) = U(z)T(y), \tag{4.4}$$

where $T(y)$ is a matrix-valued function on $\widetilde{CP^1 \backslash \mathscr{D}}$ with values in $GL(p; C)$, because for any y the images $W(y)$ and $U(z) = U(\pi(y))$ lie in the same fiber of $\widetilde{P}$.

We consider the form

$$\omega(y) = dT(y)\cdot T^{-1}(y) \tag{4.5}$$

on $\widetilde{CP^1 \backslash \mathscr{D}}$.

LEMMA 1.4.1. *The differential matrix form ω is a single-valued form on CP^1 that is holomorphic away from the points in $\mathscr{D}$, and the system* (1.1) *with the form* (4.5) *has the given monodromy* (1.5).

PROOF. We show that

$$T\circ\sigma = T\chi^{-1}(\sigma), \qquad \sigma\in\Delta \tag{4.6}$$

(recall that the group Δ is identified with $\pi_1(CP^1 \backslash \mathscr{D}, z_0))$. We have

$$(W\circ\sigma)(y) = W(\sigma y) = \langle \sigma y; I\rangle = U(z)T(\sigma y) = U(z)(T\circ\sigma)(y).$$

But on the other hand, $\langle\sigma y; I\rangle = \langle y; I\rangle\chi^{-1}(\sigma)$, since by (4.2)

$$(\sigma y; I) \sim (y; \chi^{-1}(\sigma)I) \sim (y; I)\chi^{-1}(\sigma).$$

Therefore,

$$\begin{aligned}U(z)(T\circ\sigma)(y) &= \langle\sigma y; I\rangle = \langle y; I\rangle\chi^{-1}(\sigma)\\ &= W(y)\chi^{-1}(\sigma) = U(z)T(y)\chi^{-1}(\sigma) = U(z)(T\chi^{-1}(\sigma))(y),\end{aligned}$$

which implies that $T \circ \sigma = T\chi^{-1}(\sigma)$; that is, the matrix $T(y)$ has the given monodromy (1.5).

For ω we get from (4.5) that

$$\omega(\sigma y) = dT(\sigma y)T^{-1}(\sigma y) = dT(y)\chi^{-1}(\sigma)\chi(\sigma)T^{-1}(y) = \omega(y),$$

and hence ω is "carried down" to CP^1.

In view of (4.5) the matrix-valued function $T(y)$ is a fundamental matrix of the system (1.1) with the constructed form ω, and so this system has the given monodromy (1.5).

We remark that nothing definite can be said about the character of the singularities of the system at the points $a_1, \dots, a_n$.

Let us return to the bundle $\widetilde{P}$. We extend it to a_i as follows. For $n > 2$ the inverse image $\pi^{-1}(O_i \setminus a_i)$ of the neighborhood $O_i \setminus a_i$ of a_i (see §1) consists of several connected components, of which one was singled out and denoted by $\widetilde{O}_i$. This is the component containing the terminal point y_i^0 of the lifting to $\widetilde{CP^1 \setminus \mathscr{D}}$ of the path γ_i with initial point y_0 (γ_i joins the points z_0 and $z_i^0 \in O_i \setminus a_i$). For any point $y \in \pi^{-1}(O_i \setminus a_i)$ there is a point $y_* \in \widetilde{O}_i$ such that $y = \sigma y_*$ for some $\sigma \in \Delta$.

With the help of a matrix S we reduce the matrix $E_i = \chi(g_i)$ in (1.9) to the upper triangular form $E_i^0 = S^{-1}E_iS$, where E_i^0 consists of $p_m \times p_m$ blocks $(E_i^0)^m$, $m = 1, \dots, k$, corresponding to the root subspaces ${}^mX$ of X. With each block $(E_i^0)^m$ we associate the integer vector ${}^m\varphi_i$ with coordinates

$$(4.7) \qquad {}^m\varphi_i^1 \geq \cdots \geq {}^m\varphi_i^{p_m}.$$

Denote by φ_i the vector $\varphi_i = ({}^1\varphi_i^1, \dots, {}^1\varphi_i^{p_1}, {}^2\varphi_i^1, \dots, {}^k\varphi_i^{p_k})$, and by A^{φ_i} the matrix $A^{\varphi_i} = \operatorname{diag}({}^1\varphi_i^1, \dots, {}^k\varphi_i^{p_k})$. This matrix is made up of the blocks $A^{{}^m\varphi_i} = \operatorname{diag}({}^m\varphi_i^1, \dots, {}^m\varphi_i^{p_m})$. We define the mapping

$$h^{\varphi_i}\colon \pi^{-1}(O_i \backslash a_i) \to GL(p; C)$$

as follows:

$$(4.8) \qquad h^{\varphi_i}(y) = S((z - a_i)^{A^{\varphi_i}}(y_* - a_i)^{E_i^0})S^{-1}\chi^{-1}(\sigma),$$

where $y = \sigma y_*$, $y_* \in \widetilde{O}_i$, $\sigma \in \Delta$, and $z = \pi(y) = \pi(y_*)$.

The definition (4.8) is unambiguous. Indeed, suppose that $y = \sigma' y'_*$ for some $y'_* \in \widetilde{O}_i$ and $\sigma' \in \Delta$. Then $y_* = (\sigma^{-1}\sigma')y'_*$, where $(\sigma^{-1}\sigma') = g_i^r$ for some integer r, and by virtue of (4.6), (1.8), (1.9), and (1.10′)

$$\begin{aligned} S(y_* - a_i)^{E_i^0}S^{-1}\chi^{-1}(\sigma) &= (y_* - a_i)^{E_i}\chi^{-1}(\sigma) = ((\sigma^{-1}\sigma')y'_* - a_i)^{E_i}\chi^{-1}(\sigma) \\ &= (g_i^{-r})^*(y'_* - a_i)^{E_i}\chi^{-1}(\sigma) = (y'_* - a_i)^{E_i}\chi(g_i^{-r})\chi^{-1}(\sigma) \\ &= (y'_* - a_i)^{E_i}\chi(\sigma')^{-1}\chi(\sigma)\chi^{-1}(\sigma) = (y'_* - a_i)^{E_i}\chi^{-1}(\sigma') \\ &= S(y'_* - a_i)^{E_i^0}S^{-1}\chi^{-1}(\sigma'); \end{aligned}$$

therefore,

$$h^{\varphi_i}(y) = S((z - a_i)^{A^{\varphi_i}}(y'_* - a_i)^{E_i^0})S^{-1}\chi^{-1}(\sigma'),$$

which concludes the proof that (4.8) is correct.

The mapping (4.8) has the property that

$$h^{\varphi_i}(\tau y) = h^{\varphi_i}(y)\chi^{-1}(\tau) \tag{4.9}$$

for any $\tau \in \Delta$. Indeed,

$$h^{\varphi_i}(\tau y) = S((z-a_i)^{A^{\varphi_i}}(y_* - a_i)^{E_i^0})S^{-1}\chi^{-1}(\tau\sigma) = h^{\varphi_i}(y)\chi^{-1}(\tau),$$

since $\tau y = \tau(\sigma y_*) = (\tau\sigma)y_*$.

We consider the mapping

$$\begin{gathered}\tilde{\gamma}_i\colon \tilde{\pi}_{\widetilde{P}}^{-1}(O_i\backslash a_i) \to O_i \times GL(p; C),\\ \tilde{\gamma}_i(\langle y; G\rangle) = (\pi(y); h^{\varphi_i}(y)G),\end{gathered} \tag{4.10}$$

where $\langle y; G\rangle$ is a point in $\tilde{\pi}_{\widetilde{P}}^{-1}(O_i \setminus a_i) \subset \widetilde{P}_E$ (see (4.2)).

We verify that the definition (4.10) is unambiguous. Let $(y; G)$ and $(y'; G')$ be two representatives of an equivalence class with respect to the relation $\sim$ in (4.2). Then $y' = \tau y$ and $G' = \chi(\tau)G$ for some $\tau \in \Delta$. According to (4.9),

$$h^{\varphi_i}(y')G' = h^{\varphi_i}(\tau y)G' = h^{\varphi_i}(y)\chi^{-1}(\tau)\chi(\tau)G = h^{\varphi_i}(y)G;$$

therefore,

$$(\pi(y); h^{\varphi_i}(y)G) = (\pi(y'); h^{\varphi_i}(y')G').$$

We consider the space

$$\widetilde{P}_E \cup_{\tilde{\gamma}_i} \{O_i \times GL(p; C)\}, \tag{4.11}$$

obtained by gluing the product $O_i \times GL(p; C)$ to $\widetilde{P}_E$ according to the mapping $\tilde{\gamma}_i$; that is,

$$\widetilde{P}_E \cup_{\tilde{\gamma}_i} \{O_i \times GL(p; C)\} = \widetilde{P}_E \cup \{O_i \times GL(p; C)\}/\sim,$$

where $x \sim x'$ if $\tilde{\gamma}_i(x) = x'$, $x \in \tilde{\pi}_{\widetilde{P}}^{-1}(O_i \setminus a_i)$, and $x' \in O_i \times GL(p; C)$.

We go through this procedure for all i, and denote the space obtained by P_E^{φ}, $\varphi = (\varphi_1, \dots, \varphi_n)$. Let the mapping $\pi_P\colon P_E \to CP^1$ be defined by

$$\pi_P|_{\widetilde{P}_E} = \tilde{\pi}_{\widetilde{P}}, \qquad \pi_P(y) = z, \qquad y = (z; G) \in O_j \times GL(p; C), \quad j = 1, \dots, n.$$

The bundle $P^{\varphi} = (P_E^{\varphi}, CP^1, \pi_P)$ is an extension of $\widetilde{P}$ to the whole of CP^1.

REMARK 1.4.1. The holomorphy type of P^{φ} depends in general on the way in which the matrix E_i is reduced to block upper triangular form, so that it would be more correct to write $P_{\{S\}}^{\varphi}$, but we do not do this, considering either all possible extensions $P_{\{S\}}^{\varphi}$ for different $\{S\}$, or some single extension (for some choice of $\{S\}$), depending on the situation; it will always be clear what we mean by P^{φ}. But if $\varphi = 0$ (that is, $\varphi_i = (0, \dots, 0)$, $i = 1, \dots, n$), then P^0 is clearly independent of $\{S\}$.

DEFINITION 1.4.1. An n-tuple $\varphi = (\varphi_1, \dots, \varphi_n)$ satisfying the conditions (4.7) for each i and each m will be called *admissible*.

THEOREM 1.4.1. *The RH problem for a representation* (1.5) *has a positive solution if and only if for some admissible n-tuple* $\varphi = (\varphi_1, \dots, \varphi_n)$ *the bundle* P^φ *constructed from* (1.5) *and the n-tuple* φ *with the help of* (4.11), (4.10), *and* (4.8) *is holomorphically trivial.*

PROOF. *Necessity.* Let (1.1) be a Fuchsian system with the given monodromy (1.5) and with solution space X, and let $T(y)$ be a fundamental matrix of this system with the same monodromy; that is, $T(y)$ satisfies (4.6). Let $T_i(y_*)$ be the fundamental matrix of $X|_{\widetilde{O}_i}$ constructed with respect to some Levelt basis (e). Then

$$T(y) = T(\sigma y_*) = T(y_*)\chi^{-1}(\sigma) = T_i(y_*)S^{-1}\chi^{-1}(\sigma),$$

where $y \in \widetilde{\pi}_{\widetilde{P}}^{-1}(O_i \setminus a_i)$, $y_* \in \widetilde{O}_i$, $\sigma \in \Delta$, and S^{-1} is the transition matrix from the basis (e) to the basis consisting of the columns of $T(y_*)$. We now obtain from Proposition 1.2.2 that

$$T(y) = U_i(z)(z - a_i)^{A^{\varphi_i}}(y_* - a_i)^{E_i^0}S^{-1}\chi^{-1}(\sigma), \tag{4.12}$$

where $U_i(z)$ is a holomorphically invertible matrix on O_i, and the φ_i form an admissible tuple.

We consider a bundle P^φ with the admissible n-tuple $\varphi = (\varphi_1, \dots, \varphi_n)$, where φ_i is from (4.12), $i = 1, \dots, n$. We show that this bundle is holomorphically trivial.

The mapping $U(y) = W(y)T^{-1}(y)$ in (4.4) depends only on $z = \pi(y)$ (a proof of this is obtained by formally reversing direction in the first part of the proof of Lemma 1.4.1 and using the equality (4.6)) and is a section of P^φ that is holomorphic away from the points in $\mathscr{D}$. We show that $U(z)$ is holomorphic also at these points.

From (4.11), (4.10), (4.8), and (4.12) we get that for points z in O_i

$$\begin{aligned}U(z) &= W(y)T^{-1}(y) = \langle y; I\rangle T^{-1}(y) = (\pi(y); h^{\varphi_i}(y))T^{-1}(y)\\ &= (z; S(z - a_i)^{A^{\varphi_i}}(y_* - a_i)^{E_i^0}S^{-1}\chi^{-1}(\sigma))\\ &\quad \times \chi(\sigma)S(y_* - a_i)^{-E_i^0}(z - a_i)^{-A^{\varphi_i}}U_i^{-1}(z) = (z; SU_i^{-1}(z));\end{aligned}$$

that is, $U(z)$ is a section that is holomorphic at the points a_i.

Sufficiency. Suppose that for some admissible n-tuple $\varphi = (\varphi_1, \dots, \varphi_n)$ the bundle P^φ is holomorphically trivial, and let $U(z)$ be a holomorphic section of this bundle. On $\widetilde{CP^1 \setminus \mathscr{D}}$ we consider the matrix-valued function $T(y)$ defined by (4.4). It follows from Lemma 1.4.1 that the system (1.1) with the form ω in (4.5) has the given monodromy (1.5), and its singular points are the points in $\mathscr{D}$. We show that this system is Fuchsian at the points a_i, $i = 1, \dots, n$.

Just as in the proof of necessity, we get for the points $y \in \pi_{P^\varphi}^{-1}(O_i \setminus a_i)$ that

$$U(z) = (z; S(z - a_i)^{A^{\varphi_i}}(y_* - a_i)^{E_i^0}S^{-1}\chi^{-1}(\sigma))T^{-1}(y);$$

therefore, the function

$$V_i(z) = S(z - a_i)^{A^{\varphi_i}}(y_* - a_i)^{E_i^0}S^{-1}\chi^{-1}(\sigma)T^{-1}(y)$$

is holomorphically invertible on O_i. This means that

$$T(y) = (V_i^{-1}S)(z - a_i)^{A^{\varphi_i}}(y_* - a_i)^{E_i^0}S^{-1}\chi^{-1}(\sigma),$$

where $U_i = V_i^{-1}S$ is a holomorphically invertible matrix on O_i. We get finally that the fundamental matrix $T_i(y_*) = T(\sigma^{-1}y)S$ of the space $X|_{\widetilde{O}_i}$ has the form (2.19), and $\det U_i(a_i) \neq 0$. Since φ_i is admissible and since each block $(E_i^0)^m$ of the matrix E_i^0 is upper triangular, we get as in Lemma 1.2.1 that the matrix $(z-a_i)^{A^{\varphi_i}} E_i^0 (z-a_i)^{-A^{\varphi_i}}$ is holomorphic in O_i. The fact that the system is Fuchsian now follows from Corollary 1.2.2.

REMARK 1.4.2. Since the first Chern class $c_1(P^\varphi)$ of the bundle P^φ is equal to

$$c_1(P^\varphi) = \sum_{i=1}^{n} \operatorname{tr}(A^{\varphi_i} + E_i^0) \tag{4.13}$$

($H^2(CP^1; \mathbb{Z})$ is identified with $\mathbb{Z}$), we can confine ourselves in Theorem 1.4.1 to those bundles for which $\sum_{i=1}^n \operatorname{tr}(A^{\varphi_i} + E_i^0) = 0$.

We consider the vector bundle F^φ associated with P^φ; P^φ is holomorphically trivial if and only if F^φ is. According to the Birkhoff–Grothendieck theorem (see [20]),

$$F^\varphi \cong O(-c_1^\varphi) \oplus \cdots \oplus O(-c_p^\varphi), \tag{4.14}$$

where $c_1^\varphi \geq \cdots \geq c_p^\varphi$, the c_i^φ are integers, and the bundle $O(l)$ admits the following coordinate description:

$$U_0 = CP^1 \backslash \infty, \quad U_\infty = CP^1 \backslash 0, \quad g_{0\infty} = z^l.$$

The numbers $-c_1^\varphi, \ldots, -c_p^\varphi$ are called the *partial indices* of the bundle F^φ.

DEFINITION 1.4.2. The number

$$\gamma(\varphi) = \sum_{i=1}^{p} (c_1^\varphi - c_i^\varphi)$$

is called the *weight* of the bundle F^φ.

LEMMA 1.4.2. *If for some admissible tuple φ the weight $\gamma(\varphi)$ of F^φ is zero, then the RH problem for the representation* (1.5) *has a positive solution.*

PROOF. It follows from (4.14) that F^φ is biholomorphically equivalent to the bundle given by the two neighborhoods $O_\infty = CP^1 \setminus a_1$ and $O_1 = CP^1 \setminus \infty$ and by the transition function $\gamma_{\infty 1} = (z-a_1)^{C^\varphi}$, where $C^\varphi = \operatorname{diag}(c_1^\varphi, \ldots, c_p^\varphi)$. Any other bundle biholomorphically equivalent to the given one can be specified by the same neighborhoods and the transition function

$$\tilde{\gamma}_{\infty 1} = \widetilde{U}(z)(z-a_1)^{C^\varphi}\widetilde{V}^{-1}(z),$$

where $\widetilde{U}(z)$ and $\widetilde{V}(z)$ are holomorphically invertible in O_∞ and O_1, respectively. From the last remark and the fact that the restrictions of the bundle P^φ to O_1 and O_∞ are holomorphically trivial it follows that P^φ has a section $U(z)$ that is holomorphic away from a_1 and has the form $U(z) = V(z)(z-a_1)^{-C^\varphi}$ in a neighborhood of this point, where $V(z)$ is a holomorphic section of P^φ over a neighborhood of a_1. In this case the matrix-valued function $T(y)$ in (4.4) takes the form

$$T(y) = (z-a_1)^{C^\varphi}\widetilde{V}^{-1}(z)S(z-a_1)^{A^{\varphi_1}}(y_*-a_1)^{E_i^0}S^{-1}\chi^{-1}(\sigma)$$

instead of (4.12). It follows from the condition $\gamma(\varphi) = 0$ that C^{φ} is the scalar matrix $c_1^{\varphi} I$; therefore, the expression for T can be rewritten as

$$T(y) = \widetilde{V}^{-1}(z)S(z-a_1)^{A^{\varphi_1}+c_1^{\varphi}I}(y_* - a_1)^{E_i^0}S^{-1}\chi^{-1}(\sigma). \tag{4.15}$$

We consider the bundle $F^{\varphi'}$, where $\varphi_i' = \varphi_i$ for $i \neq 1$, and $\varphi_1' = \varphi_1 + (c_1^{\varphi}, \dots, c_1^{\varphi})$. The tuple φ' is admissible, and the bundle $F^{\varphi'}$ is holomorphically trivial in view of (4.15). The lemma now follows from Theorem 1.4.1.

Let us now consider the canonical extension F^0 of the bundle $\widetilde{F}$ (for $\varphi = 0$).

THEOREM 1.4.2. *Suppose that each of the monodromy matrices $G_i = \chi(g_i)$ of (1.5) reduces to a Jordan block. If the representation (1.5) is reducible, then the RH problem for this representation has a positive solution if and only if $\gamma(0) = 0$.*

PROOF. The sufficiency was proved in Lemma 1.4.2.

Necessity. We consider a Fuchsian system (1.1) with monodromy (1.5) and with solution space X. It follows from the reducibility of the monodromy that there is a subspace $X_l \subset X$ of dimension $0 < l < p$ that is invariant under the action of the monodromy operators g^*. We consider a basis (e) of $X|_{\widetilde{O}_i}$ in which the matrix of g_i^* is a Jordan block. The only l-dimensional subspace of $X|_{\widetilde{O}_i}$ invariant under the action of g_i^* is the space spanned by the first l vectors of (e) (the space Y^l of the filtration (1.19)). Therefore $X_l|_{\widetilde{O}_i} = Y^l$, and hence the first l normalizations of X at a_i that are maximal in size are the normalizations of X_l at this point.

Indeed, in Example 1.1.3 it was proved that the Jordan basis (e) is an associated basis of the space $X|_{\widetilde{O}_i}$, and this, in particular, means that

$$\varphi_i^1 \geq \cdots \geq \varphi_i^p, \qquad \operatorname{Re}\beta_i^1 \geq \cdots \geq \operatorname{Re}\beta_i^p, \tag{4.16}$$

where $\varphi_i^j = \varphi_i(e_j)$; but $(e_1, \dots, e_l)$ is a basis of the subspace $X_l|_{\widetilde{O}_i}$ by what was proved.

Denote by Σ_l the sum of the exponents of the subspace X_l, and by Σ the sum of the exponents of X over all the points a_i. From (4.16),

$$\Sigma_l \geq \frac{1}{p}\Sigma, \tag{4.17}$$

with equality only when

$$\varphi_i^1 = \cdots = \varphi_i^p, \qquad i = 1, \dots, n. \tag{4.18}$$

But from Lemma 1.3.7 and Proposition 1.2.3 we get that $0 \geq \Sigma_l \geq \frac{l}{p}\Sigma = 0$, whence $\Sigma_l = \Sigma$, and (4.18) holds.

Applying the $\Gamma(z) = \prod_{i=2}^{n} B(a_i, a_1, \varphi_i^1, \dots, \varphi_i^1)$ procedure to the space X, we get the space $X' = \Gamma(z)X$ with normalizations $\{'\varphi_i^j\}$ such that

$$'\varphi_i^1 = \cdots = {}'\varphi_i^p = 0, \quad i \neq 1, \qquad {}'\varphi_1^1 = \cdots = {}'\varphi_1^p \tag{4.19}$$

Again denote it by X, and its normalizations by $\{\varphi_i^j\}$. The decompositions (2.6) for a space $X|_{\widetilde{O}_i}$ with $i \neq 1$ will have the form

$$T_i(y) = U_i(z)(y - a_i)^{E_i}, \tag{4.20}$$

while at a_1

$$T_1(y) = U_1(z)(z-a_1)^{\varphi_1^1 I}(y-a_1)^{E_1}, \tag{4.21}$$

where the matrices $U_i(z)$, $i=1,\dots,n$, coincide with the analogous matrices in (2.6) for the original space X. Therefore, $\det U_i(a_i)\neq 0$, $i=1,\dots,n$, and X is again (by Proposition 1.2.3) the solution space of some Fuchsian system (and with the same monodromy (1.5) as before).

We consider the bundle P^0 constructed from (1.5), and the section

$$U(z) = W(y)T^{-1}(y) \tag{4.22}$$

of this bundle (see (4.4)), where $T(y)$ is a fundamental matrix of the space X. Just as in the proof of necessity in Theorem 1.4.1, we get from (4.20) and (4.21) that $U(z)$ is a holomorphic section of the bundle P^φ, where $\varphi=(\varphi_1, 0, \dots, 0)$ and $\varphi_1=(\varphi_1^1, \dots, \varphi_1^1)$. Since the restrictions of P^0 and P^φ to the complement of the neighborhood O_1 of a_1 coincide, $U(z)$ is a section of P^0 that is holomorphic outside O_1. It follows from (4.21) that in O_1

$$U(z) = V(z)(z-a_1)^{-\varphi_1^1 I}, \tag{4.23}$$

where $V(z)$ is a holomorphic section of the bundle $P^0|_{O_1}$. We get immediately from (4.23) that the vector bundle F^0 associated with P^0 is holomorphically equivalent to the bundle given by the two neighborhoods $O_\infty = CP^1\backslash a_1$ and O_1 and the transition function $\gamma_{\infty 1}=(z-a_1)^{\varphi_1^1 I}$, and this last bundle (hence also F^0) is holomorphically equivalent to the direct sum of bundles

$$O(-\varphi_1^1)\oplus\cdots\oplus O(-\varphi_1^1)\cong F^0. \tag{4.24}$$

Therefore, $\gamma(0)=n(\varphi_1^1-\varphi_1^1)=0$. The theorem is proved.

REMARK 1.4.3. If we omit the requirement that the representation (1.5) be reducible, then the condition $\gamma(0)=0$ is no longer be necessary for solvability of the RH problem (see Example 2.2.1).

COROLLARY 1.4.1. *Suppose that the representation* (1.5) *is reducible and that each of the monodromy matrices* $G_i=\chi(g_i)$ *reduces to a Jordan block. If the RH problem for* (1.5) *has a positive solution, then each Fuchsian system* (1.1) *with the given monodromy* (1.5) *satisfies the conditions* (4.18). *In addition, there exists a Fuchsian system* (1.1) *with the given monodromy such that the normalizations* $\{\varphi_i^j\}$ *of its solution space satisfy the conditions* (4.19):

$$\varphi_i^1=\cdots=\varphi_i^p=0,\quad i\neq l,\qquad \varphi_l^1=\cdots=\varphi_l^p, \tag{4.19$'$}$$

where $1\le l\le n$ *is an arbitrary previously specified number.*

If X_i *is a subspace of* X *invariant under the action of the monodromy operators, then the sum* Σ_i *of its exponents is equal to zero.*

The next assertion was proved by Plemelj in [52]. It also follows easily from the constructions in this section.

PROPOSITION 1.4.1. *For any representation* (1.5) *there exists a system* (1.1) *with regular singular points* $a_1, \ldots, a_n$ *and with the given monodromy* (1.5) *that is Fuchsian at* $a_1, \ldots, \hat{l}, \ldots, a_n$ *and such that all the normalizations* φ_i^j *of the solution space* X *of this system at the points* $a_1, \ldots, \hat{l}, \ldots, a_n$ *are equal to zero.*

PROOF. From the representation (1.5) we construct the bundle F^0 with admissible tuple $\varphi = (0, \ldots, 0)$. It follows from the decomposition (4.14) that this bundle has a meromorphic section $U(z)$ that is holomorphic away from some previously specified point (for example, a_l). We consider the matrix-valued function $T(y)$ in (4.4) and the system (1.1) with the matrix form ω in (4.5). Since $U(z)$ is meromorphic, $T(y)$ has finite order at a_l. It follows from the proof of Theorem 1.4.1 and Lemma 1.4.1 that the system is Fuchsian at $a_1, \ldots, \hat{l}, \ldots, a_n$ and has the given monodromy (1.5). The proof of the same Theorem 1.4.1 (formula (4.12)) tells us that all the normalizations of the solution space of this system are equal to 0 at all the points a_i, $i \neq l$.

5. Examples of the positive solution of the Riemann-Hilbert problem

In this section we present some examples of the positive solution of the RH problem, examples that can easily be obtained by the methods developed above.

PROPOSITION 1.5.1. *For a commutative representation* (1.5) *the RH problem has a positive solution.*

PROOF. We consider the matrices E_i in (1.9). It follows from the circuit relation $G_1 \cdot \ldots \cdot G_n = I$ and the commutativity of the representation that the matrices E_i commute pairwise, and

(5.1) $$E_1 + \cdots + E_n = A,$$

where A is a diagonal integer matrix.

Consider the matrix-valued function

(5.2) $$T(y) = (z - a_1)^{-A} \prod_{i=1}^{n} (y - a_i)^{E_i}.$$

It follows from the commutativity of (1.5) and from (1.3) and (1.8) that $T(y)$ has the given monodromy (1.5) and is a fundamental matrix of the solution space of the system (1.1) with matrix form

(5.3) $$\omega = dT \cdot T^{-1} = \left(\frac{E_1 - A}{z - a_1} + \sum_{i=2}^{n} \frac{E_i}{z - a_i} \right) dz.$$

We note that $\operatorname{res}_\infty \omega = A - \sum_{i=1}^{n} E_i = 0$; therefore, the system (1.1) with matrix form ω is Fuchsian on CP^1, and its singular points are the given points $a_1, \ldots, a_n$.

The next assertion was proved in [2].

PROPOSITION 1.5.2. *Suppose that the representation* (1.5) *is such that one of the monodromy matrices is diagonalizable. Then the RH problem for* (1.5) *has a positive solution.*

PROOF. It can be assumed without loss of generality that the monodromy matrix G_1 is diagonalizable. Consider a system (1.1) with regular singular points $a_1, \ldots, a_n$

and monodromy (1.5) that is Fuchsian at $a_2, \ldots, a_n$. The existence of such a system was proved in Proposition 1.4.1. Denote by X the solution space of this system, and consider a fundamental matrix $T(y)$ of the space $X|_{\tilde{O}_1}$ such that in the basis of its columns the monodromy operator g_1^* has diagonal form.

We represent $T(y)$ in the form (2.6):

$$T(y) = U(z)(z-a_1)^A(y-a_1)^{E_1}, \tag{5.4}$$

where A is a diagonal integer matrix and $U(z)$ is a holomorphic matrix in the neighborhood O_1 of a_1 (in particular, A can be the zero matrix).

We show that there exists a meromorphic matrix $\Gamma(z)$ on CP^1 that is holomorphically invertible away from a_1 and such that

$$\Gamma(z)U(z) = V(z)(z-a_1)^B, \tag{5.5}$$

where B is a diagonal integer matrix and $V(z)$ is a holomorphically invertible matrix on O_1.

Since the order of the zero of $\det U(z)$ at a_1 is finite, the proof of the last assertion follows immediately from the next lemma.

LEMMA 1.5.1. *For any matrix $U(z)$ that is holomorphically invertible in $O_1 \setminus a_1$, holomorphic at a_1, and satisfies $\det U(a_1) = 0$ there is a matrix $\Gamma(z)$ that is meromorphic on CP^1, holomorphically invertible away from a_1, and such that*

$$\Gamma(z)U(z) = U_1(z)(z-a_1)^{B_1}, \tag{5.6}$$

where $U_1(z)$ is holomorphic in O_1, and B_1 is a diagonal integer matrix with nonnegative elements on the main diagonal and positive trace: $\operatorname{tr} B_1 > 0$.

PROOF. Since $\det U(a_1) = 0$, by a linear combination of rows (that is, by multiplying $T(y)$ from the left by a nonsingular constant matrix S) we can reduce $U(z)$ to a matrix $U' = SU$ for which some row is of the form (3.6) with a number $k > 0$. We apply the $A(a_1, l, t, k)$ procedure to U' and get the matrix $U_1 = \Gamma_1(z)U'$, which in view of Lemma 1.3.2 can be represented in the form (5.6) with the matrix $B_1 = \operatorname{diag}(0, \ldots, 0, k, 0, \ldots, 0)$ (k in the tth place). Let $\Gamma(z) = \Gamma_1(z)S$. The lemma is proved.

We pass from the system (1.1) constructed above to a system with solution space $X' = \Gamma(z)X$, where $\Gamma(z)$ is from (5.5). Since $\Gamma(z)$ is holomorphically invertible away from a_1, this system will again be Fuchsian at $a_2, \ldots, a_n$, as follows from (3.2). We get from (5.4) and (5.5) that the fundamental matrix $T'(y) = \Gamma(z)T(y)$ of this system has in a neighborhood of a_1 the form

$$T'(y) = V(z)(z-a_1)^{A+B}(y-a_1)^{E_1},$$

where the matrix $V(z)$ is holomorphically invertible in the neighborhood O_1, and $A+B$ is a diagonal integer matrix.

Since the diagonal matrices $A+B$ and E_1 commute, Corollary 1.2.2 now tells us that the system constructed is Fuchsian at a_1.

The next result was actually obtained by Dekkers in [41]. Below we present a different, simplified variant of the proof.

PROPOSITION 1.5.3. *For a representation* (1.5) *of dimension* $p = 2$ *the RH problem has a positive solution.*

PROOF. If the representation (1.5) of dimension $p = 2$ is reducible, then either all the monodromy matrices G_i can be reduced simultaneously to Jordan blocks, and hence the representation is commutative, or one of the matrices G_i is diagonalizable. In either case we get from Propositions 1.5.1 and 1.5.2 that for a reducible representation (1.5) of dimension $p = 2$ the RH problem has a positive solution.

Suppose that (1.5) is irreducible. We consider a system (1.1) with regular singular points $a_1, \dots, a_n$ and monodromy (1.5) that is Fuchsian away from a_1 (the existence of such a system was proved in Proposition 1.4.1); denote its solution space by X. We choose an associated basis (e_1, e_2) in $X|_{\widetilde{O}_1}$ and consider the decomposition (2.6):

$$T(y) = U_1(z)(z - a_1)^{A_1}(y - a_1)^{E_1}. \tag{2.6}$$

Suppose that $\det U_1(z) = (z - a_1)^k(c + o(1))$, with $c \neq 0$ and $k > 0$. Then by Proposition 1.2.3 the sum Σ of the exponents of X is less than zero: $\Sigma < 0$. By the same proposition, to prove the current proposition it suffices to show that there is always a meromorphic matrix $\Gamma(z)$ on CP^1 that is holomorphically invertible away from a_1 such that the sum $\widetilde{\Sigma}$ of the exponents of the space $\widetilde{X} = \Gamma(z)X$ satisfies the inequality

$$\widetilde{\Sigma} > \Sigma. \tag{5.7}$$

By a linear combination of rows (that is, by multiplying from the left by a nonsingular matrix S) we can reduce the matrix $U_1(z)$ in (2.6) to a matrix $U_1' = SU_1$ whose second row has the form

$$u_2'(z) = ((z - a_1)^{k_1}v_1(z), (z - a_1)^{k_2}v_2(z)), \tag{5.8}$$

where $k_1 > 0$, $k_2 > 0$, $v_1(z)$ and $v_2(z)$ are holomorphic functions on O_1, and $v_1(a_1) \neq 0$.

The last inequality actually does hold, for if $v_1(a_1) = 0$ for any $k_1 > 0$, then the element $u_{21}'(z)$ of U_1' is identically equal to zero, and hence the representation (1.5) is reducible by virtue of Lemma 1.3.6, and this is false.

We apply the $A(a_1, 2, 1, k_1)$ procedure to the matrix $T'(y)$ and get the solution space $\widetilde{X} = \Gamma_1(z)X' = \Gamma_1 SX$ of a system (1.1) with regular singular points $a_1, \dots, a_n$. It follows from Lemma 1.3.3 that the system constructed is again Fuchsian at $a_2, \dots, a_n$. From Remark 1.3.1 we get that the sum $\widetilde{\Sigma}$ of the exponents of $\widetilde{X}$ satisfies the inequality (5.7): $\widetilde{\Sigma} = \Sigma + k_1 - \max(0, k_1 - k_2) > \Sigma$.

The proposition is proved.

CHAPTER II

The Riemann–Hilbert Problem for Systems of Three Equations

1. Positive solution of the Riemann–Hilbert problem for an irreducible representation of dimension three

We consider the representation (1.1.5).([1])

THEOREM 2.1.1. *For any irreducible representation* (1.1.5) *of dimension* $p = 3$ *the RH problem has a positive solution.*

PROOF. Consider the solution space X of a system of three equations (1.1.1) with regular singular points $a_1, \dots, a_n$ that is Fuchsian at $a_2, \dots, a_n$ and has irreducible monodromy representation (1.1.5). (The existence of such a system was proved in Proposition 1.4.1.) By the criterion (1.2.22) for being Fuchsian and by the condition (1.2.21), to prove the theorem it suffices to show that there exists a meromorphic matrix-valued function $\Gamma(z)$ on CP^1 that is holomorphically invertible away from the points $a_1, \dots, a_n$ and such that the following conditions hold for the space $X' = \Gamma(z)X$:

a) the space X' is the solution space of a system (1.1.1) that is Fuchsian at all but perhaps one of the points $a_1, \dots, a_n$;

b) the inequality

$$\sum_{i=1}^{n}\sum_{j=1}^{3} {}'\beta_i^j > \sum_{i=1}^{n}\sum_{j=1}^{3} \beta_i^j \tag{1.1}$$

holds, where β_i^j and ${}'\beta_i^j$ are the exponents of the respective spaces X and X'.

Consider the decomposition (1.2.6) for the space $X|_{\widetilde{O}_1}$. We pass from the space X to $X' = \Gamma(z)X$, where $\Gamma(z)$ is the matrix determined in Lemma 1.3.9. Denote by $U_1'(z)$ the matrix

$$U_1'(z) = \Gamma(z)U_1(z) = (z - a_1)^C V_1(z). \tag{1.2}$$

The rest of the proof breaks up into three cases:

1) $c_{11} > 0$;

2) $c_{11} = 0,\ c_{22} > 0$;

([1])Here and below, when referring to a formula in another chapter, we give the chapter number before the formula number.

3) $c_{11} = c_{22} = 0$, $c_{33} > 0$.

We treat these cases in succession.

1. In the first case the definition of normalizations and the form of the decomposition (1.2.6) for the space $X'|_{\tilde{O}_1}$ give us that $'\varphi_1^j \geq \varphi_1^j + c_{11}$, $j = 1, 2, 3$, for the normalizations φ_1^j and $'\varphi_1^j$ of the respective spaces X and X'. We get the conditions (1.1) a) and b) from the last inequalities.

2. Denote X' again by X, and U_1' by U_1. It follows from the irreducibility of the representation (1.1.5) and Lemma 1.3.6 that either $v_{l1}(A_1) \neq 0$, where $l = 2$ or $l = 3$ for the elements of the matrix $V_1(z)$ in (1.2), or there are numbers $2 \leq i \leq n$ and $2 \leq l \leq n$ such that the element $u_{l1}(z)$ of the matrix $U_i(z)$ in the decomposition (1.2.6) of the space $X|_{\tilde{O}_i}$ is nonzero at a_i: $u_{l1}(a_i) \neq 0$. In the first case the lth row of $U_1(z)$ has the form (1.3.10) with $t = 1$. Applying to X the $A(a_1, l, 1, c_{ll})$ procedure in Lemma 1.3.3, we get that the conditions (1.1) a) and b) are satisfied for the space $X' = \Gamma(z)X$. But if $u_{l1}(a_i) \neq 0$, then we apply the $B(a_1, a_i, 0, c_{22}, c_{33})$ procedure to X and get a space $X' = \Gamma X$ that is the solution space of a system (1.1.1) that is Fuchsian at $a_1, \ldots, \hat{\imath}, \ldots, a_n$. The row u_l' of the holomorphic matrix $U_i' = \Gamma U_i$ will have the form (1.3.10) with $t = 1$ and $k = c_{ll} > 0$. Applying the $A(a_i, l, 1, c_{ll})$ procedure to X', we again get the conditions (1.1) a) and b) for the space $X'' = \Gamma' X'$.

3. Let us again denote X' by X and U_i' by U_i. We consider the elements u_{lm}^i, $1 \leq l, m \leq 3$, of the matrix U_i in the decomposition (1.2.6) for the space $X|_{\tilde{O}_i}$, and the elements v_{lm} of the matrix $V_1(z)$ in (1.2). The third case breaks up into two subcases:

a) one of the numbers $v_{31}(a_1), u_{31}^2(a_2), \ldots, u_{31}^n(a_n)$ is nonzero;

b) the functions $v_{31}(z), u_{31}^2(z), \ldots, u_{31}^n(z)$ have zeros of orders $k_1, \ldots, k_n$ at the respective points $a_1, \ldots, a_n$, and $k_i > 0$, $i = 1, \ldots, n$.

Case 3a) was analyzed in 2). Let us consider case 3b). It follows from the irreducibility of the representation (1.1.5) and Lemma 1.3.4 that the numbers $k_1, \ldots, k_n$ are finite. Denote by s the sum

$$s = k_1 + c_{33} + k_2 + \cdots + k_n \tag{1.3}$$

of the orders of the zeros of the functions $u_{31}^i(z)$ at the points a_i, $i = 1, \ldots, n$. It also follows from the irreducibility of (1.1.5) and Lemma 1.3.6 that either $v_{32}(a_1) \neq 0$, or there is an i such that $u_{32}^1(a_i) \neq 0$. In the latter case we apply to X the $B(a_1, a_i, 0, 0, c_{33})$ procedure and get a space $X' = \Gamma(z)X$ that is the solution space of a system (1.1.1) that is Fuchsian at $a_1, \ldots, \hat{\imath}, \ldots, a_n$. Here the sum s in (1.3) does not change. Denote this space again by X. The matrix $U_i(z)$ in the decomposition (1.2.6) for this space has the form (1.3.26), where $C = \operatorname{diag}(0, 0, c_{33})$.

The rest of the proof of the theorem is based on a certain procedure that we call the L procedure, and describe below.

Description of the L procedure. It follows from Remark 1.2.1 that either $u_{11}^i(a_i) \neq 0$ or $u_{21}^i(a_i) \neq 0$. We pass from X to $X' = SX$, where S is a nonsingular constant matrix of the form

$$S = \begin{pmatrix} \multicolumn{2}{c}{S'} & 0 \\ & & 0 \\ 0 & 0 & 1 \end{pmatrix} \tag{1.4}$$

and such that the first column x_1 of the matrix $U_i'(a_i) = SU_i(a_i)$ has the form

$$x_1 = (1, 0, 0)^T. \tag{1.5}$$

Applying the $A(a_i, 3, 2, c_{33})$ procedure to the space X', we get a space ${}^1X = \Gamma_1 X'$ such that the matrix ${}^1U_i(z) = \Gamma_1 U_i'(z)$ has the form

$${}^1U_i(a_i) = \begin{pmatrix} 1 & 0 & * \\ 0 & 0 & * \\ 0 & 0 & 0 \end{pmatrix}. \tag{1.6}$$

It follows from Lemma 1.3.6 that there is a j_1 such that the value ${}^1u_{21}^j(a_j)$ of the element ${}^1u_{21}^j(z)$ of ${}^1U_j(z)$ at a_j is nonzero: ${}^1u_{21}^j(a_j) \neq 0$. Applying the $A(a_j, 2, 1, 0)$ procedure to ${}^1X$, we get a space ${}^2X = \Gamma_2(z)^1X$. It follows from the form of the transformations used that the matrices ${}^2U_m(z) = \Gamma_2(z)^1U_m(z)$ satisfy the following conditions:

i) the third row of ${}^2U_m(z)$ coincides with the third rows of the matrices $U_m(z)$, $m = 1, \dots, n$;

ii) the matrix ${}^2U_i(a_i)$ has the form (1.6);

iii) the first column of the matrix ${}^2U_j(a_j)$ has the form $(0, {}^1u_{21}^j(a_j), 0)^T$.

We apply the $B(a_i, a_j, 0, 1, 0)$ procedure to the space ${}^2X$ and denote the resulting space by $\widetilde{X}$. This completes the description of the L procedure.

It follows from the form (1.6) of the matrix ${}^2U_i(a_i)$ that

$$\tilde{\varphi}_i^1 = \varphi_i^1, \quad \tilde{\varphi}_i^2 \geq \varphi_i^2, \quad \tilde{\varphi}_i^3 \geq \varphi_i^3 - 1, \tag{1.7}$$

where φ_i^j and $\tilde{\varphi}_i^j$ are the normalizations of the spaces X and $\widetilde{X}$, respectively. The form iii) of the first column of ${}^2U_j(a_j)$ implies that

$$\tilde{\varphi}_j^1 = \varphi_j^1 + 1, \quad \tilde{\varphi}_j^2 = \varphi_j^2, \quad \tilde{\varphi}_j^3 = \varphi_j^3, \tag{1.8}$$

and the decomposition (1.2.6) for the space $X|_{\widetilde{O}_j}$ has the form

$$\widetilde{T}_j(y) = \widetilde{U}_j'(z)(z - a_j)^{A_j + N}(y - a_j)^{E_j}, \tag{1.9}$$

where $N = \operatorname{diag}(1, 0, 0)$, and $\widetilde{U}_j'(z) = \widetilde{U}_j(z)(z - a_j)^{-N}$. We get from i) and (1.9) that the sums s and $\tilde{s}$ in (1.3) for the spaces X and $\widetilde{X}$ are connected by the relation

$$\tilde{s} = s - 1 < s. \tag{1.10}$$

It follows from (1.7) and (1.8) that the space $\widetilde{X}$ satisfies the condition (1.1) a), and

$$\sum_{k=1}^{n} \sum_{m=1}^{3} \tilde{\beta}_k^m \geq \sum_{k=1}^{n} \sum_{m=1}^{3} \beta_k^m. \tag{1.11}$$

Let us return to the proof of the theorem. We apply the L procedure to X. If $\widetilde{X}$ satisfies the condition 3b), then we again apply the L procedure. It follows from (1.10) that in at most $\sum_{l=1,\, l\neq i}^{n}(k_l - 1) + 1$ steps we get a space satisfying 3a). By

(1.11), the proof of the theorem in this case again reduces to case 2, already analyzed. The theorem is proved.

2. Fuchsian systems of two equations on the Riemann sphere. The Fuchsian weight of a representation

We consider a Fuchsian system (1.1.1) of two equations on CP^1 with singular points $a_1, \dots, a_n$. It can be assumed without loss of generality that the point ∞ is not a singular point of this system (this condition can always be achieved by a conformal transformation of the Riemann sphere). In this case the form ω of the system (1.1.1) can be written as

$$\omega = \sum_{i=1}^{n} B^i \frac{dz}{z - a_i}, \tag{2.1}$$

where $B^i = \operatorname{res}_{a_i} \omega$, and the residue theorem gives us that

$$\sum_{i=1}^{n} B^i = 0 \tag{2.2}$$

(see the Introduction). We consider the solution space X of the system (1.1.1), (2.1), along with Levelt bases for each $X|_{\widetilde{O}_i}$ and corresponding decompositions (1.2.6) for all $i = 1, \dots, n$. Then we can assume that the normalizations φ_i^1 and φ_i^2 of X at a_i satisfy

$$\varphi_i^1 \geq \varphi_i^2, \qquad i = 1, \dots, n. \tag{2.3}$$

Indeed, if the matrix G_i of the operator g_i^* is a Jordan block, then (2.3) is automatically satisfied; but if G_i is a diagonal matrix and $\varphi_i^1 < \varphi_i^2$, then it suffices to permute the elements of the Levelt basis, obtaining again a Levelt basis for which (2.3) now holds.

DEFINITION 2.2.1. The *Fuchsian weight* γ_ω of a Fuchsian system (1.1.1) of two equations is defined to be the number

$$\gamma_\omega = \sum_{i=1}^{n} (\varphi_i^1 - \varphi_i^2).$$

Consider the set Ω of all Fuchsian systems of two equations on CP^1 with the given monodromy representation (1.1.5) (by Proposition 1.5.3, this set is nonempty).

DEFINITION 2.2.2. The *Fuchsian weight* γ_χ of the representation (1.1.5) is defined to be the number

$$\gamma_\chi = \min_{\Omega} \gamma_\omega .$$

PROPOSITION 2.2.1. *For any representation* (1.1.5) *of dimension* $p = 2$

$$\gamma_\omega \geq 0, \qquad \gamma_\chi \geq 0, \tag{2.4}$$

and the parity of the numbers γ_ω *and* γ_χ *is the same as the parity of the number* $\sum_{i=1}^n \operatorname{tr} E_i$. *If* (1.1.5) *is a commutative representation that does not decompose into a direct sum of one-dimensional representations, then* $\gamma_\chi = 0$.

PROOF. The first part of the proposition follows from (2.3) and the fact that

$$\sum_{i=1}^{n}(\beta_i^1+\beta_i^2)=\sum_{i=1}^{n}\operatorname{tr}E_i+\sum_{i=1}^{n}(\varphi_i^1+\varphi_i^2).$$

Indeed, we get from (1.2.22) that

$$\sum_{i=1}^{n}(\varphi_i^1+\varphi_i^2)=-\sum_{i=1}^{n}\operatorname{tr}E_i,$$

and hence

$$\gamma_\omega=\sum_{i=1}^{n}(\varphi_i^1-\varphi_i^2)=-\sum_{i=1}^{n}\operatorname{tr}E_i-2\sum_{i=1}^{n}\varphi_i^2.$$

If (1.1.5) is a commutative representation that does not decompose into a direct sum of one-dimensional representations, then the matrices G_i can be simultaneously reduced to the form

$$\begin{pmatrix}\lambda_i & *\\ 0 & \lambda_i\end{pmatrix}.$$

Therefore, the matrix

$$T(y)=(z-a_1)^{-\frac{1}{2}\sum_{i=1}^{n}\operatorname{tr}E_i}\prod_{i=1}^{n}(y-a_i)^{E_i} \tag{2.5}$$

is a fundamental matrix of a Fuchsian system (1.1.1) with the given monodromy and weight 0.

LEMMA 2.2.1. *For any Fuchsian system* (1.1.1) *with Fuchsian weight* γ_ω *and with solution space* X *there exist a point* $a_l\in\mathscr{D}$ *and a meromorphic matrix-valued function* $\Gamma(z)$ *on* CP^1 *that is holomorphically invertible away from the points in* $\mathscr{D}$ *and such that the space* $X'=\Gamma(z)X$ *satisfies the following conditions*:

a) X' *is the solution space of some Fuchsian system* (1.1.1);

b) $'\varphi_i^1={}'\varphi_i^2=0$ *for* $i=1,\dots,\hat{l},\dots,n$, *and* $'\varphi_l^1-{}'\varphi_l^2\le\gamma_\omega$, *where* $\{'\varphi_i^j\}$ *is the set of normalizations of* X'.

PROOF. At each point a_i we consider a decomposition (1.2.6) for X satisfying the condition (2.3). Denote by J the set of indices for which $\varphi_i^1>\varphi_i^2$, $i\in J$. If this set is empty or $J=\{l\}$ for some l, then we apply to X the transformation $\prod_{i\neq l}B(a_i,a_l,\varphi_i^1,\varphi_i^1)$ (the index l is chosen arbitrarily in the case $J=\varnothing$). The lemma follows easily from the form of the decompositions (1.2.6) in this case.

Suppose that the set J contains elements $i, m, \dots$. To prove the lemma it suffices to show that by means of a substitution $\widetilde{X}=\Gamma X$ it is possible to reduce the number of elements in the set J without increasing the Fuchsian weight of the system.

We pass from the space X to $X'=U_i^{-1}(a_i)X$, which we again denote by X. In the decomposition (1.2.6) for X we now have

$$U_i(a_i)=I. \tag{2.6}$$

Applying the $B(a_i,a_m,1,0)$ procedure to X, we obtain a space $X'=\Gamma_1(z)X$. It follows from the form of the matrix $\Gamma_1(z)U_i(z)$ that the normalizations $'\varphi_i^j$ of X'

are connected with the normalizations φ_i^j of X by the relations

$$'\varphi_i^1 = \varphi_i^1 - 1, \qquad '\varphi_i^2 = \varphi_i^2, \qquad \Delta_i' = '\varphi_i^1 - '\varphi_i^2 = \Delta_i - 1, \tag{2.7}$$

where $\Delta_i = \varphi_i^1 - \varphi_i^2$.

We consider the matrix $U_m(z)$ in the decomposition (1.2.6) for X. The rest of the proof breaks up into two cases: a) $u_{11}^m(a_m) \neq 0$; b) $u_{11}^m(a_m) = 0$.

In case a) the first row of $\Gamma_1(z)U_m(z)$ has the form (1.3.10) with $t = 1$ and $k = 1$. We apply the $A(a_m, 1, 1, 1)$ procedure to X' and get the space $X'' = \Gamma_2 X'$, whose normalizations $''\varphi_m^j$ at a_m satisfy in view of Lemma 1.3.3 the relations

$$''\varphi_m^1 = \varphi_m^1 + 1, \qquad ''\varphi_m^2 = \varphi_m^2, \qquad \Delta_m'' = \Delta_m + 1. \tag{2.8}$$

In case b) it follows from Proposition 1.2.2 that $u_{12}^m(a_m) \neq 0$, and hence the first row of $\Gamma_1(z)U_m(z)$ has the form (1.3.10) with $t = 2$ and $k = 1$. Applying the $A(a_m, 1, 2, 1)$ procedure to X', we get a space $X'' = \Gamma_2(z)X'$ whose normalizations $''\varphi_m^j$ satisfy in view of Lemma 1.3.3 the relations

$$''\varphi_m^1 = \varphi_m^1, \qquad ''\varphi_m^2 = \varphi_m^2 + 1, \qquad \Delta_m'' = \Delta_m - 1. \tag{2.9}$$

It follows from (2.7)–(2.9) that for the space X''

$$\sum_{l=1}^{n} (''\varphi_l^1 - ''\varphi_l^2) \leq \gamma_\omega, \qquad \Delta_i'' = \Delta_i - 1 \tag{2.10}$$

both in case a) and in case b).

If $\Delta_i'' > 0$ and $\Delta_m'' > 0$, we repeat the procedure once more. It follows from (2.10) that in at most $\Delta_i = \varphi_i^1 - \varphi_i^2$ steps we get a space $\widetilde{X} = \Gamma(z)X$ for which (2.10) holds and either $\widetilde{\Delta}_i = 0$ or $\widetilde{\Delta}_m = 0$. The latter means that the number of elements in J has been reduced. We get from (2.7)–(2.10) and Proposition 1.2.3 that $\widetilde{X}$ is the solution space of a Fuchsian system (1.1.1) whose Fuchsian weight does not exceed γ_ω.

REMARK 2.2.1. In general, for $n > 3$ an arbitrary Fuchsian system (1.1.1) cannot be reduced to the form a), b) of Lemma 2.2.1 at an arbitrary previously specified point $a_l \in \mathscr{D}$.

COROLLARY 2.2.1. *Any Fuchsian system of two equations* (1.1.1) *with Fuchsian weight* $\gamma_{\omega'}$ *can be reduced by a transformation* (1.3.1) *to a Fuchsian system* (1.1.1) *with Fuchsian weight* γ_ω *not exceeding* $\gamma_{\omega'}$ *such that the decompositions* (1.2.6) *for its solution space* X *have the form*

$$T_i(y) = U_i(z)(y - a_i)^{E_i}, \qquad i \neq l, \tag{2.11}$$

$$T_l(y) = U_l(z)(z - a_l)^{A_l}(y - a_l)^{E_l}, \tag{2.12}$$

where all the matrices $U_j(z)$ *are holomorphically invertible at the points* a_j, *the* E_i

are upper triangular matrices,

$$E_l = \begin{pmatrix} \rho_l^1 & \varepsilon \\ 0 & \rho_l^2 \end{pmatrix}, \tag{2.13}$$

$$A_l = \begin{pmatrix} b+\gamma_1 & 0 \\ 0 & b-\gamma_2 \end{pmatrix}, \tag{2.14}$$

$$U_l(z) = \begin{pmatrix} 1 & c(z-a_l)^m \\ s(z-a_l)^k & 1 \end{pmatrix} (1+o(1)), \tag{2.15}$$

where $\gamma_1 = [(\gamma_\omega + 1)/2]$, $\gamma_1 + \gamma_2 = \gamma_\omega$, $m, k > 0$, $c \neq 0$ *if* $u_{12}^l(z) \not\equiv 0$, $s \neq 0$ *if* $u_{21}^l(z) \not\equiv 0$, *and* $[x]$ *is the integer part of* x.

PROOF. The decompositions (2.11) and (2.12) are direct consequences of Lemma 2.2.1. In (2.13) we have that $\varepsilon \neq 0$ if E_l is a Jordan block, and $\varepsilon = 0$ otherwise. To get (2.15) we must pass from the space $\widetilde{X}$ constructed in Lemma 2.2.1 to the space $\widehat{X} = \widetilde{U}_l^{-1}(a_l)\widetilde{X}$ and the corresponding system.

LEMMA 2.2.2. *Suppose that the Fuchsian system* (1.1.1) *is reduced to the form* (2.11)–(2.15), *and* $\gamma_\omega > 0$. *If all the monodromy matrices* G_j *of this system are nondiagonalizable, then there is an* $i \neq l$ *such that the value of the element* $u_{11}^i(z)$ *of the matrix* $U_i(z)$ *in the decomposition* (2.11) *at* a_i *is nonzero:*

$$u_{11}^i(a_i) \neq 0. \tag{2.16}$$

PROOF. By assumption, all the matrices E_j in (2.11) and (2.12) are Jordan blocks. Assume that $u_{11}^i(a_i) = 0$ for $i = 1, \dots, \hat{l}, \dots, n$. Then from (2.1), (1.2.11), and (2.11) we get that

$$B^i = U_i(a_i)E_iU_i^{-1}(a_i) = \begin{pmatrix} \rho_i & 0 \\ * & \rho_i \end{pmatrix}.$$

It follows from (2.15), (1.2.11), and $\gamma_\omega > 0$ that $B^l = \operatorname{diag}(\beta_l^1, \beta_l^2)$. Therefore, by (2.2),

$$\sum_{\substack{i=1\\ i\neq l}}^n \rho_i + \beta_l^1 = \sum_{\substack{i=1\\ i\neq l}}^n \rho_i + \beta_l^2 = 0,$$

from which $\beta_l^1 = \beta_l^2$, and since $\rho_l^1 = \rho_l^2$, we have $\varphi_l^1 = \varphi_l^2$, which contradicts the condition $\gamma_\omega > 0$.

LEMMA 2.2.3. *Suppose that a Fuchsian system* (1.1.1) *with monodromy* (1.1.5) *is reduced to the form* (2.11)–(2.15), *and the numbers* c *and* m *in* (2.15) *are such that* $c \neq 0$ *and* $m \leq \frac{1}{2}\gamma_\omega$. *Then* $\gamma_\chi < \gamma_\omega$.

PROOF. Applying the $A(a_l, 1, 2, m)$ procedure to the solution space X of (1.1.1), we obtain a space $X' = \Gamma(z)X$. It follows from Lemma 1.3.2 and the form (1.3.8) of the matrix $\Gamma(z)$ that the matrix

$$U_l' = \Gamma(z)U_l(z)(z-a_l)^C, \tag{2.17}$$

where $C = \operatorname{diag}(m, -m)$, is holomorphically invertible at a_l. Since $\varphi_l^1 - m \geq \varphi_l^2 + m$ by a condition of the lemma, the decomposition (1.2.6) for $X'|_{\widetilde{O}_l}$ takes the form

$$T_l'(y) = U_l'(z)(z-a_l)^{A_l'}(y-a_l)^{E_l}, \tag{2.18}$$

where $A_l' = A_l - C$, and hence $\gamma_\chi \leq \gamma_{\omega'} = \gamma_\omega - 2m$.

In some cases the Fuchsian weight of a representation (1.1.5) can be determined from the Fuchsian weight of any system with the monodromy (1.1.5).

PROPOSITION 2.2.2. *Suppose that a Fuchsian system* (1.1.1) *with monodromy* (1.1.5) *is reduced to the form* (2.11)–(2.15). *If the number* c *in* (2.15) *is equal to zero, then* $\gamma_\chi = \gamma_\omega$. *If* $c \neq 0$, *then* $\gamma_\chi = \gamma_\omega$ *if and only if* $m \geq \gamma_\omega$.

PROOF. 1. We prove first that if $c \neq 0$ and $m < \gamma_\omega$, then $\gamma_\chi < \gamma_\omega$. By Lemma 2.2.3, it suffices to analyze the case

$$\tfrac{1}{2}\gamma_\omega < m < \gamma_\omega. \tag{2.19}$$

There are two possibilities for the matrix E_l in (2.13): a) E_l is diagonal; b) E_l is a Jordan block.

In case a) we apply the $A(a_l, 1, 2, m)$ procedure to X and get a space $X' = \Gamma(z)X$ for which we have the decomposition (2.17), (2.18) and ${}'\varphi_l^1 = \varphi_l^1 - m$, ${}'\varphi_l^2 = \varphi_l^2 + m$. Permuting the columns of the matrix $T_l' = \Gamma T_l$, we get a decomposition for X' of the form (1.2.6) (since E_l is a diagonal matrix) with the matrix $A_l'' = \operatorname{diag}(\varphi_l^2 + m, \varphi_l^1 - m)$, $E_l'' = \operatorname{diag}(\rho_l^2, \rho_l^1)$. From (2.19) it follows that

$$\gamma_{\omega'} = \varphi_l^2 + m - (\varphi_l^1 - m) = 2m - \gamma_\omega < \gamma_\omega, \tag{2.20}$$

and therefore, $\gamma_\chi < \gamma_\omega$.

Let us consider case b). If E_i is a diagonal matrix for some $i \neq l$, then by permuting the columns of $T_i(y)$ if necessary, we get that the (2.16) holds for the $U_i(z)$ in (2.11). If all the E_j with $j \neq l$ are Jordan blocks, then in view of Lemma 2.2.2 there is an $i \neq l$ for which (2.16) holds.

Applying the $B(a_l, a_i, 2m - \gamma_\omega, 0)$ procedure to X, we get the space $X' = \Gamma_1(z)X$. The matrix

$$U_l'(z) = \Gamma_1(z)U_l(z - a_l)^C,$$

where $C = \operatorname{diag}(2m - \gamma_\omega, 0)$, is holomorphically invertible at a_l, its first row has the form

$$((a_l - a_i)^{2m-\gamma_\omega}, \quad c'(z - a_l)^{\gamma_\omega - m})(1 + o(1)), \qquad c' \neq 0,$$

and the normalizations ${}'\varphi_l^j$ of X' at a_l are equal to ${}'\varphi_l^1 = \varphi_l^1 - (2m - \gamma_\omega)$, ${}'\varphi_l^2 = \varphi_l^2$. Note that ${}'\varphi_l^1 - {}'\varphi_l^2 = 2(\gamma_\omega - m)$; therefore, by applying the $A(a_l, 1, 2, \gamma_\omega - m)$ procedure to X', we get a space $X'' = \Gamma_2(z)X'$ whose normalizations ${}''\varphi_l^j$ are equal:

$${}''\varphi_l^1 = {}''\varphi_l^2 \tag{2.21}$$

according to Lemma 2.2.3.

The $A(a_l, 1, 2, \gamma_\omega - m)$ procedure does not change the form (1.3.10) of the first row of the matrix $\Gamma_1(z)U_i(z)$ with $t = 1$ and $k = 2m - \gamma_\omega$. Applying the $A(a_i, 1, 1, 2m - \gamma_\omega)$ procedure to X'', we get a space $\widetilde{X} = \Gamma_3(z)X''$ whose normalizations $\widetilde{\varphi}_i^j$ satisfy

$$\widetilde{\varphi}_i^1 = \varphi_i^1 + 2m - \gamma_\omega, \qquad \widetilde{\varphi}_i^2 = \varphi_i^2 \tag{2.22}$$

in view of Lemma 1.3.3 and (2.20). From (2.21), (2.22), and (2.19) we get

$$\gamma_{\tilde{\omega}} = \sum_{j=1}^{n}(\tilde{\varphi}_j^1 - \tilde{\varphi}_j^2) = 2m - \gamma_\omega < \gamma_\omega$$

for the space $\widetilde{X}$. Hence, $\gamma_\chi < \gamma_\omega$.

2. We show that if $c \neq 0$ and $m \geq \gamma_\omega$, or if $c = 0$, then $\gamma_\chi = \gamma_\omega$. Assume the contrary. Suppose that there exists a Fuchsian system (1.1.1) with solution space X' with the same monodromy as the original system and such that its Fuchsian weight is equal to $\gamma_{\omega'} < \gamma_\omega$. According to Corollary 2.2.1, it can be assumed that this system is reducible to the form (2.11)–(2.15) with a_l replaced by a_t and the numbers γ_1, γ_2 replaced by γ_1', γ_2'. From the assumption that $\gamma_{\omega'} < \gamma_\omega$ and the fact that $\gamma_{\omega'}$ and γ_ω have the same parity (see Proposition 2.2.1) it follows that

$$\gamma_1' < \gamma_1, \qquad \gamma_2' < \gamma_2. \tag{2.23}$$

In the space X' we choose a basis (e') such that for the fundamental matrix $T'(y)$ constructed with respect to this basis and for the matrix $T_l(y)$ of X in (2.12) we have

$$(T')^{-1}(g_i^* T')(y) = T_l^{-1}(g_i^* T_l)(y). \tag{2.24}$$

Let us consider $(e')|_{\widetilde{O}_t}$. There are two possibilities for the normalizations φ_t of the basis (e'):

a) $\varphi_t(e_1') \neq \varphi_t(e_2')$;

b) $\varphi_t(e_1') = \varphi_t(e_2') = b - \gamma_2'$.

In case a) the basis (e_1', e_2') is weakly associated. In case b) there exist $\lambda_1 \neq 0$ and $\lambda_2 \neq 0$ such that

$$\varphi_t(\lambda_1 e_1' + \lambda_2 e_2') = b + \gamma_1'.$$

We pass from the base (e) of X and the base (e') of X' to $(\tilde{e}) = (e)S$ and $(\tilde{e}') = (e')S$, where

$$S = \begin{pmatrix} 1 & \lambda_1/\lambda_2 \\ 0 & 1 \end{pmatrix} \tag{2.25}$$

in case b), and $S = I$ in case a).

In case b) we have $\varphi_t(\tilde{e}_1') = b - \gamma_2'$ and $\varphi_t(\tilde{e}_2') = b + \gamma_1'$ by construction, and hence the basis $(\tilde{e}_1', \tilde{e}_2')$ is weakly associated.

Thus, for the fundamental matrix $\widetilde{T}' = T'S$, where S is from (2.25), we have the decomposition (1.2.6), (1.2.8) with the matrix

$$\widetilde{A}_t' = \operatorname{diag}(b_{11}, b_{22}), \tag{2.26}$$

where one of the numbers b_{jj} is equal to $b + \gamma_1'$, and the other is equal to $b - \gamma_2'$.

For the matrix $\widetilde{T}_l = T_l S$, considered on $\widetilde{O}_l$, we again have the decomposition (2.12)–(2.15) with the matrix $U_l(z)$ replaced by $\widetilde{U}_l(z)$ and with E_l replaced by $S^{-1}E_l S$, where

$$\widetilde{U}_l(z)(z-a_l)^{A_l} = U_l(z)(z-a_l)^{A_l}S = U_l(z)\widetilde{S}(z)(z-a_l)^{A_l},$$

$$\widetilde{S}(z) = \begin{pmatrix} 1 & (\lambda_1/\lambda_2)(z-a_l)^{\gamma_\omega} \\ 0 & 1 \end{pmatrix}$$

in case b), and $\widetilde{S} = I$ in case a). Therefore, in case b)

$$\widetilde{U}_l(z) = \begin{pmatrix} 1 & c(z-a_l)^m + (\lambda_1/\lambda_2)(z-a_l)^{\gamma_\omega} \\ s(z-a_l)^k & 1 \end{pmatrix}(1+o(1)).$$

By assumption either $c \neq 0$ and $m \geq \gamma_\omega$, or $c = 0$, so $\widetilde{U}_l(z)$ again has the form (2.15) with c replaced by c' and m by m', and either $c' \neq 0$ and $m' \geq \gamma_\omega$, or $c' = 0$.

Consider the matrix-valued function

$$Y(y) = \widetilde{T}_l(y)(\widetilde{T}'(y))^{-1}. \tag{2.27}$$

It follows from (2.24) and (2.25) that $Y(y)$ is a single-valued function on CP^1. From (2.11) we get that $Y(z)$ is holomorphically invertible away from the points a_l and a_t, and from (2.12)–(2.15) we get that in a neighborhood of a_l

$$Y(z) = \widetilde{U}_l(z)(z-a_l)^{A_l}(\widetilde{U}'_1)^{-1};$$

therefore, the first row $y_1(z)$ of Y can be represented there in the form

$$y_1(z) = ((z-a_l)^{b+\gamma_1}, c'(z-a_l)^{b+m'-\gamma_2})((\widetilde{U}'_l(a_l))^{-1} + o(1)),$$

from which, since $m' \geq \gamma_\omega$ if $c' \neq 0$ (and in the case $c' = 0$, of course), we get

$$y_1(z) = (z-a_l)^{b+\gamma_1} t_1(z), \tag{2.28}$$

where $t_1(z)$ is a holomorphic row vector at a_l.

In a neighborhood of a_t the matrix Y has the form

$$Y(z) = \widetilde{U}_t(z)(z-a_t)^{-A'_t}(\widetilde{U}'_t(z))^{-1},$$

where A'_t has the form (2.26). In this case we get for the same row of Y that

$$y_1(z) = \frac{1}{(z-a_t)^{b+\gamma'_1}} t_2(z), \tag{2.29}$$

where $t_2(z)$ is a holomorphic vector at a_t.

It follows from (2.23), (2.28), (2.29), and the holomorphicity of $y_1(z)$ away from the points a_l and a_t that the degree of the divisor of each component of $y_1(z)$ on CP^1 is strictly greater than zero, and hence $y_1(z) \equiv 0$, which contradicts the holomorphic invertibility of $Y(z)$ away from a_l and a_t. This contradiction shows that $\gamma_\chi = \gamma_\omega$.

LEMMA 2.2.4. *Suppose that a Fuchsian system* (1.1.1), (2.1) *is reduced to the form* (2.11)–(2.15). *Then in a neighborhood of a_l the form ω of this system is*

$$\omega = \begin{pmatrix} \beta_1^1/(z-a_l) & c(m-\beta_1^1+\beta_1^2)(z-a_l)^{m-1} + \varepsilon(z-a_l)^{\gamma_m-1} \\ s(k+\beta_1^1-\beta_1^2)(z-a_l)^{k-1} & \beta_1^2/(z-a_l) \end{pmatrix} \times (1+o(1))\,dz. \tag{2.30}$$

PROOF. From (2.15),

$$U_l^{-1} = \begin{pmatrix} 1 & -c(z-a_l)^m \\ -s(z-a_l)^k & 1 \end{pmatrix} (1+o(1)),$$
$$\frac{dU_l}{dz} = \begin{pmatrix} (z-a_l)^{t_1} & cm(z-a_l)^{m-1} \\ sk(z-a_l)^{k-1} & (z-a_l)^{t_2} \end{pmatrix} (1+o(1)),$$

where $t_1 \geq 0$ and $t_2 \geq 0$. The relation (2.30) now follows from (1.2.9).

LEMMA 2.2.5. *Suppose that the element* $\omega_{pq} dz$ *of the matrix form* ω *of a Fuchsian system* (1.1.1), (2.1) *has the form*

$$\omega_{pq} = \frac{1}{z-a_l} c_{-1}^{pq} + c_0^{pq} + (z-a_l)c_1^{pq} + \cdots \tag{2.31}$$

in a neighborhood of a_l. *The elements* b_{pq}^1 *of the matrix* B^i *in* (2.1) *are connected with the numbers* c_k^{pq} *by the relations*

$$b_{pq}^{l} = c_{-1}^{pq}, \qquad \sum_{\substack{i=1\\ i\neq l}}^{n} b_{pq}^{i} \frac{1}{(a_i-a_l)^r} = -c_{r-1}^{pq}, \qquad r = 1, \ldots \tag{2.32}$$

PROOF. From (2.2),

$$\omega - B^l \frac{dz}{z-a_l} = \sum_{\substack{i=1\\ i\neq l}}^{n} B^i \frac{dz}{z-a_i},$$

which yields the proof.

PROPOSITION 2.2.3. *The Fuchsian weight* γ_χ *of any representation* (1.1.5) *satisfies the inequality* $\gamma_\chi \leq n-1$. *If* (1.1.5) *is irreducible, then*

$$\gamma_\chi \leq n-2. \tag{2.33}$$

PROOF. Consider a Fuchsian system (1.1.1), (2.1) with monodromy (1.1.5) and with Fuchsian weight γ_χ, and let it be reduced to the form (2.11)–(2.15). Assume that $\gamma_\chi > n-2$. Then $m \geq n-1$ for the number m in (2.15), according to Proposition 2.2.2. In this case it follows from Lemma 2.2.4 that the element ω_{12} of the form ω of this system has the form

$$\omega_{12} = o((z-a_l)^{n-3})dz$$

in a neighborhood of a_l. Therefore, we get from Lemma 2.5 that the numbers b_{12}^i with $i \neq l$ satisfy the system of $n-2$ equations (2.32):

$$\sum_{\substack{i=1\\ i\neq l}}^{n} b_{12}^{i} \frac{1}{(a_i-a_l)^r} = 0, \qquad r = 1, \ldots, n-2. \tag{2.34}$$

It follows from the condition $\gamma_\chi > 0$ and (1.2.11) that $b_{12}^l = 0$, and from (2.2) that $b_{12}^1 + \cdots + b_{12}^n = 0$. Thus, the equation

$$\sum_{\substack{i=1\\ i\neq l}}^{n} b_{12}^{i} = 0$$

can be adjoined to the system (2.34).

We obtain a system whose determinant is the Vandermonde determinant of the numbers $1/(a_i - a_l), \dots, \hat{l}, \dots, 1/(a_n - a_l)$, and is thus nonzero. Hence, $b^i_{12} = 0$, $i = 1, \dots, n$; therefore, all the matrices B^i of the system (1.1.1), (2.1) have lower triangular form, and we get from Corollary 1.2.3 that

$$b^l_{11} = \rho^1_l + \varphi^1_l, \quad b^l_{22} = \rho^2_l + \varphi^2_l, \quad b^i_{11} = \rho^1_i, \quad b^i_{22} = \rho^2_i, \quad i \neq l. \tag{2.35}$$

Since (2.2) implies that $\sum_{i=1}^n \rho^j_i + \varphi^j_l = 0$ for $j = 1, 2$, and (1.1.9) implies that $0 \leq \operatorname{Re} \sum_{i=1}^n \rho^j_i \leq n-1$ for $j = 1, 2$, we have from the last two equalities that $-(n-1) \leq \varphi^j_l \leq 0$ for $j = 1, 2$, and therefore $\varphi^1_l - \varphi^2_l \leq n-1$. Together with the assumption $\gamma_\chi > n-2$, this leads to the equality $\gamma_\chi = n-1$.

In this case the fact that the matrices B^i are lower triangular gives us that the representation (1.1.5) is reducible.

PROPOSITION 2.2.4. *For any points $a_1, \dots, a_n$ and any number γ satisfying the inequalities*

$$0 < \gamma \leq n-2, \tag{2.36}$$

there exists an irreducible representation (1.1.5) *such that $\gamma_\chi = \gamma$, and all the monodromy matrices G_i of the group* $\operatorname{Im}\chi$ *in* (1.1.9) *are nondiagonalizable.*

PROOF. Suppose that $a_1 = 0$ and that the point ∞ is not in $\mathscr{D}$ (this can always be achieved by a conformal transformation of the Riemann sphere).

1. We prove the proposition first for the case of even $\gamma = 2\gamma'$. Consider the following two systems of equations in the unknowns $d_2, \dots, d_n$ and $c_2, \dots, c_n$:

$$\sum_{i=2}^n d_i \frac{1}{a_i^r} = \delta_{r,\gamma}, \tag{2.37}$$

$$\sum_{i=2}^n c_i \frac{1}{a_i^r} = x^2 \delta_{r,n-2}, \tag{2.38}$$

where $r = 0, \dots, n-2$, and $\delta_{i,j}$ is the Kronecker symbol.

Since the determinant of the system (2.37), (2.38) is nonzero, for the solution $\mathbf{d}$ of (2.37) there is a j such that $d_j \neq 0$. And every solution $\mathbf{c}$ of (2.38) has the form $\mathbf{c} = x^2\mathbf{t}$, where $t_i \neq 0$ for $i = 2, \dots, n$; therefore, there are values of the square roots $\sqrt{d_i t_i}$ such that $s = \sum_{i=2}^n \sqrt{d_i t_i} \neq 0$. Let the number x in (2.38) be chosen equal to $x = -\gamma/2s = -\gamma'/s$. Then $\sum_{i=2}^n \sqrt{d_i c_i} = xs = -\gamma/2 = -\gamma'$, and hence the matrices

$$B^1 = \begin{pmatrix} \gamma' & 0 \\ 0 & -\gamma' \end{pmatrix}, \quad B^i = \begin{pmatrix} \sqrt{d_i c_i} & d_i \\ -c_i & -\sqrt{d_i c_i} \end{pmatrix}, \quad i = 1, \dots, n, \tag{2.39}$$

satisfy the conditions (2.2).

Let us consider a Fuchsian system (1.1.1), (2.1), (2.39). From Corollary 1.2.3 and the form of the matrices (2.39) we get for the exponents of the solution space X of this system that

$$\beta^1_1 = \varphi^1_1 = \gamma', \qquad \beta^2_1 = \varphi^2_1 - \gamma', \qquad \beta^j_i = \rho^j_i = \varphi^j_i = 0, \quad i \neq 1, \ j = 1, 2. \tag{2.40}$$

We consider the decomposition (1.2.6) for the space $X|_{\widetilde{O}_1}$:

$$T_1(y) = U_1(z)z^{A_1}y^{E_1}. \tag{2.41}$$

From (2.40),

$$A_1 = \begin{pmatrix} \gamma' & 0 \\ 0 & -\gamma' \end{pmatrix}, \qquad E_1 = \begin{pmatrix} 0 & \alpha \\ 0 & 0 \end{pmatrix}, \tag{2.42}$$

where $\alpha = 0$ or $\alpha = 1$. We get from (2.39) and (1.2.11) that

$$\begin{pmatrix} \gamma' & 0 \\ 0 & -\gamma' \end{pmatrix} = U_1(0)\begin{pmatrix} \gamma' & 0 \\ 0 & -\gamma' \end{pmatrix}U_1^{-1}(0),$$

and thus

$$U_1(0) = \begin{pmatrix} u_{11} & 0 \\ 0 & u_{22} \end{pmatrix}.$$

In the space $X|_{\widetilde{O}_1}$ we make the change of basis $e_1' = e_1/u_{11}$, $e_2' = e_2/u_{22}$. In the new basis, which we denote again by (e), the matrices in (2.41) take the form (2.15), (2.42) with a_l replaced by $a_1 = 0$ and α by $\varepsilon = \alpha u_{11}/u_{22}$. From (2.37) and Lemma 2.2.5 we get for the matrix form ω of the constructed system (1.1.1), (2.1), (2.39) that

$$\omega_{12} = (z^{\gamma-1} + o(z^{\gamma-1}))\,dz. \tag{2.43}$$

It follows from (2.43) and Lemma 2.2.4 that the number m in (2.15) for the matrix $U_1(z)$ satisfies the inequality $m \geq \gamma$, and we get from this inequality and Proposition 2.2.2 that for the constructed system $\gamma_\omega = \gamma_\chi = \gamma$.

It remains to show that the monodromy matrices $G_1, \dots, G_n$ of the constructed system are nondiagonalizable, and its monodromy representation is irreducible.

To show that $G_1, \dots, G_n$ are nondiagonalizable it suffices in view of (1.1.9) to show that $E_1, \dots, E_n$ are nondiagonalizable. It follows from (2.40) and (1.2.11) that the matrices L_i with $i \neq 1$ coincide with E_i. Therefore, in view of (1.2.11) the matrices E_i with $i \neq 1$ are nondiagonalizable because the B^i in (2.39) are nondiagonalizable.

We show that E_1 is also nondiagonalizable. Note that for $c \neq 0$ and $m = \gamma$, and for $c = 0$, the expression $c(m - \beta_1^1 + \beta_1^2)$ in (2.30) is zero in view of (2.40), and hence the element ω_{12} in (2.30) has for $m \geq \gamma$ the form

$$\omega_{12} = (\varepsilon z^{\gamma_\omega - 1} + o(z^{\gamma_\omega - 1}))\,dz.$$

Comparing this with (2.43) and using the equality $\gamma_\omega = \gamma_\chi = \gamma$ already proved, we get that $\varepsilon = 1$ and $\alpha = u_{22}/u_{11} \neq 0$. Therefore, E_1 is nondiagonalizable.

If the monodromy representation of the system (1.1.1), (2.1), (2.39) were reducible, it would follow from the nondiagonalizability of the matrices $G_1, \dots, G_n$ that it is commutative. In this case Proposition 2.2.1 would give us that $\gamma_\chi = 0$, which contradicts the proven equality $\gamma_\chi = \gamma > 0$.

2. We consider the case of odd $\gamma = 2\gamma'' + 1$. The proof of the proposition in this case repeats in entirety the case of even γ with the matrices B^i in (2.39) replaced

by

(2.39′)
$$B^1 = \begin{pmatrix} -n+2+\gamma'' & 0 \\ 0 & -n+1-\gamma' \end{pmatrix}, \quad B^i = \begin{pmatrix} \sqrt{d_ic_i} + \frac{2n-3}{2n-2} & d_i \\ -c_i & -\sqrt{d_ic_i} + \frac{2n-3}{2n-2} \end{pmatrix},$$

the matrix A_1 replaced by

$$A_1 = \begin{pmatrix} -n+2+\gamma'' & 0 \\ 0 & -n+1-\gamma'' \end{pmatrix}$$

and the exponents in (2.40) replaced by

$$\text{(2.40}') \qquad \begin{aligned} &\beta_1^1 = \varphi_1^1 = -n+2+\gamma'', \qquad \beta_1^2 = \varphi_1^2 = -n+1-\gamma'' \\ &\beta_i^j = \rho_i^j = \frac{2n-3}{2n-2}, \qquad i \neq 1, \; j = 1, 2. \end{aligned}$$

REMARK 2.2.2. Since the matrix T_1 in (2.41) has the form (2.12)-(2.15), it follows that

$$\varphi_1^1(e_{11}^1) = \gamma', \qquad \varphi_1(e_{12}^1) \geq 0, \qquad \varphi_1(e_{21}^1) \geq \gamma', \qquad \varphi_1(e_{22}^1) = -\gamma'$$

for the elements e_{ij}^1 of T_1.

EXAMPLE 2.2.1. The monodromy representation of the Fuchsian system

$$\text{(2.44)} \qquad \begin{aligned} df = \Bigg[&\begin{pmatrix} 1 & 0 \\ 0 & -1 \end{pmatrix} \frac{dz}{z} + \frac{1}{6} \begin{pmatrix} -1 & 1 \\ -1 & 1 \end{pmatrix} \frac{dz}{z+1} \\ &+ \frac{1}{2} \begin{pmatrix} -1 & -1 \\ 1 & 1 \end{pmatrix} \frac{dz}{z-1} + \frac{1}{3} \begin{pmatrix} -1 & 1 \\ -1 & 1 \end{pmatrix} \frac{dz}{z - \frac{1}{2}} \Bigg] f \end{aligned}$$

with four singular points $a_1 = 0$, $a_2 = -1$, $a_3 = 1$, and $a_4 = 1/2$ has Fuchsian weight $\gamma_\chi = 2$, and the monodromy matrices $G_1, \dots, G_n$ are nondiagonalizable. This example will be thoroughly analyzed in §4.

3. A criterion for positive solvability of the Riemann–Hilbert problem for a representation of dimension three

In this section we describe all the representations (1.1.5) of dimension $p = 3$ for which the RH problem has a positive solution.

PROPOSITION 2.3.1. *Suppose that a representation (1.1.5) of dimension $p = 3$ is reducible, and that each of the generators G_i in (1.1.9) of the monodromy group reduces to a Jordan block. The RH problem for (1.1.5) has a positive solution if and only if the equality*

$$\text{(3.1)} \qquad \gamma_{\chi^2} = 0$$

holds for the corresponding two-dimensional subrepresentation or factor representation χ^2 of the representation (1.1.5).

PROOF. *Necessity.* By Corollary 1.4.1, the equalities (1.4.18) hold for any Fuchsian system with the monodromy (1.1.5).

We consider a fundamental matrix $T(y)$ of the solution space X of this system with the first i columns forming a basis for a space X_i that is invariant under the action of the monodromy operators $(i = 1, 2)$. Then the matrix $T'(y) = U_1^{-1}(a_1)T(y)$

has the form (1.3.19):

$$\text{a) } i=2 \qquad\qquad \text{b) } i=1$$

(3.2)
$$T'(y)=\begin{pmatrix} \widetilde{T}(y) & & \begin{matrix} * \\ * \end{matrix} \\ 0 \quad 0 & & e_{33} \end{pmatrix}, \qquad T'(y)=\begin{pmatrix} e_{11} & * \quad * \\ \begin{matrix} 0 \\ 0 \end{matrix} & \widetilde{T}(y) \end{pmatrix}.$$

Indeed, otherwise the sum Σ_i of the exponents of X_i could be increased according to the proof of Lemma 1.3.8, but $\Sigma_i = 0$ by virtue of Corollary 1.4.1; therefore, such an increase would contradict the inequality (1.3.16) in Lemma 1.3.7. In (3.2) the matrix $\widetilde{T}$ is a fundamental matrix of the solution space $\widetilde{X}$ of a Fuchsian system (1.1.1) with monodromy χ^2, and hence we get from (1.4.18) and (3.2) a) that the normalizations $\widetilde{\varphi}_j^k$ of $\widetilde{X}$ satisfy $\widetilde{\varphi}_j^1 = \varphi_j^1$ and $\widetilde{\varphi}_j^2 = \varphi_j^2$, and $\gamma_{\chi^2}^j = 0$ for $i = 2$.

By the definition of normalizations, $\widetilde{\varphi}_j^1 \geq \varphi_j^2$ and $\widetilde{\varphi}_j^2 \geq \varphi_j^3$ in the case (3.2) b), and from this we again get from $\Sigma_i = 0$ and the condition (1.2.22) for being Fuchsian that $\widetilde{\varphi}_j^1 = \varphi_j^2$ and $\widetilde{\varphi}_j^2 = \varphi_j^3$, and $\gamma_{\chi^2} = 0$ for $i = 1$.

Sufficiency. 1. Consider a system (1.1.1) with regular singular points $a_1, \dots, a_n$ that is Fuchsian at $a_2, \dots, a_n$ with monodromy (1.1.5) and with a fundamental matrix T of the solution space X of the form (3.2) (the existence of such a system was proved in Lemma 1.3.8). From Proposition 1.4.1 and Corollary 2.2.1 we get that there exists a meromorphic matrix $\widetilde{\Gamma}(z)$ on CP^1 that is holomorphically invertible away from the points $a_1, \dots, a_n$ and such that the space $\widetilde{X}'$ with fundamental matrix $\widetilde{T}' = \widetilde{\Gamma}(z)\widetilde{T}$ has decomposition (2.11)–(2.15), where $\gamma_1 = \gamma_2 = 0$ in view of the condition $\gamma_{\chi^2} = 0$.

Denote by c_j the number $\varphi_j(e_{33})$ in the case (3.2) a) and the number $\varphi_j(e_{11})$ in the case (3.2) b). We pass from the space X to $X' = \Gamma(z)X$, where

$$\text{a) } i=2 \qquad\qquad \text{b) } i=1$$

(3.3)
$$\Gamma(z)=\begin{pmatrix} \Gamma_1(z) & \begin{matrix} 0 \\ 0 \end{matrix} \\ 0 \quad 0 & 1 \end{pmatrix}, \qquad \Gamma(z)=\begin{pmatrix} 1 & 0 \quad 0 \\ \begin{matrix} 0 \\ 0 \end{matrix} & \Gamma_1(z) \end{pmatrix},$$
$$\Gamma_1(z)=\prod_{\substack{j=1\\ j\neq 1}}^{n}\left(\frac{z-a_j}{z-a_l}\right)^{c_j}\widetilde{\Gamma}(z).$$

The fundamental matrix $T' = \Gamma T$ of the space X' again has the form (3.2), and the following equalities hold for the normalizations ${}'\varphi_j^k$ of this space (for the respective cases a) and b)):

(3.4)
$$\text{a) } {}'\varphi_j^1 = {}'\varphi_j^2 = \varphi_j(e_{33}) = c_j, \qquad j = 1, \dots, n,$$
$$\text{b) } c_j = {}'\varphi_j^1 = \varphi_j(\widetilde{e}_1) = \varphi_j(\widetilde{e}_2), \qquad j = 1, \dots, n,$$

where $\widetilde{e}_1$ and $\widetilde{e}_2$ are the columns of $\widetilde{T}$. The equalities (3.4) with $j \neq l$ follow from the form of the matrices $\Gamma_1(z)$ and (2.11). We prove them for $j = l$. Denote by ρ_j the eigenvalues of the Jordan block E_j. For the functions $e_{11}(y)$ and $e_{33}(y)$ we have by the residue theorem for the form $d\ln e_{jj}(y)$ that

(3.5)
$$\sum_{j=1}^{n}(\rho_j + c_j) = 0.$$

On the other hand, for the space $\widetilde{X}'$ we get by (1.2.22) that

$$\sum_{j=1}^{n}(2\rho_j + {}'\widetilde{\varphi}_j^1 + {}'\widetilde{\varphi}_j^2) = 0. \tag{3.6}$$

Since ${}'\widetilde{\varphi}_j^1 = {}'\widetilde{\varphi}_j^2$, this and (3.5) give us that $\sum_{j=1}^n c_j = \sum_{j=1}^n {}'\widetilde{\varphi}_j^1 = \sum_{j=1}^n {}'\widetilde{\varphi}_j^2$. The case $j = l$ now follows from the equalities (3.4) with $j \neq l$.

2. The rest of the proof breaks up into two cases: a) $i = 2$; b) $i = 1$.

Consider first the case b). It follows from (3.4) and the construction of the matrix $\widetilde{T}'$ that $T'(y)$ can be represented in $\widetilde{O}_j$ in the form

$$T'(y) = U_j(z)(z - a_j)^{c_j I}(y - a_j)^{E_j}, \tag{3.7}$$

where

$$U_j(z) = \begin{pmatrix} u_{11} & u_{12} & u_{13} \\ 0 & & \\ 0 & \multicolumn{2}{c}{\widetilde{U}_j(z)} \end{pmatrix},$$

$u_{11}(a_j) \neq 0$, $\widetilde{U}_j(z)$ is holomorphically invertible at a_j, and in general $u_{12}(z)$ and $u_{13}(z)$ have poles of respective order p and q at a_j.

Denote by $u_2(z)$ and $u_3(z)$ the rows of the matrix $\widetilde{U}_j(z)$. Since $u_2(a_j)$ and $u_3(a_j)$ are linearly independent, by Proposition 1.3.1 there exist polynomials $Q_2(\frac{1}{z-a_j})$ and $Q_3(\frac{1}{z-a_j})$ of degrees at most $r = \max(p, q)$ such that the vector $t(z) = Q_2u_2 + Q_3u_3 + (u_{12}, u_{13})$ is holomorphic at a_j. (To find Q_2 and Q_3, apply Proposition 1.3.1 to the rows of the matrix $(z - a_j)^r U_j(z)$.) We pass from the space X' to ${}^1X' = \Gamma_2(z)X'$, where

$$\Gamma_2(z) = \begin{pmatrix} 1 & Q_2 & Q_3 \\ 0 & 1 & 0 \\ 0 & 0 & 1 \end{pmatrix}.$$

The matrix ${}^1U_j' = \Gamma_2(z)U_j$ will now be holomorphically invertible at a_j; therefore, the equality (3.4) b) takes the form ${}'\varphi_j^1 = {}'\varphi_j^2 = {}'\varphi_j^3 = c_j$. We follow the procedure described for all $j = 1, \dots, n$ and obtain a space ${}^nX'$. From the holomorphic invertibility of ${}^nU_j'$, the form of the decompositions (3.7), and Corollary 1.2.2 it follows that ${}^nX'$ is the solution space of a Fuchsian system (1.1.1) with the given monodromy (1.1.5).

Now consider case a). Applying the $A(a_j, 3, 3, 0)$ procedure to the matrix $U_j(z) = T'(y)(y - a_j)^{-E_j}(z - a_j)^{-c_j I}$, we get a space ${}^1X'$ for which the matrix ${}^1U_j'(z)$ in the decomposition (3.7) is now holomorphically invertible. We follow this procedure for all $j = 1, \dots, n$. The rest of the proof is a repetition of case b).

COROLLARY 2.3.1. *Suppose that a representation* (1.1.5) *of dimension* $p = 3$ *is reducible, and that each of the matrices* $G_i' = \chi^2(g_i)$ *of the corresponding two-dimensional subrepresentation or factor representation* χ^2 *reduces to the form*

$$\begin{pmatrix} \lambda_i & * \\ 0 & \lambda_i \end{pmatrix}.$$

If $\gamma_{\chi^2} = 0$, *then the RH problem for the representation* (1.1.5) *has a positive solution.*

PROOF. If each of the monodromy matrices G_i has an eigenvalue of multiplicity three, then the proof is the same as in Proposition 2.3.1. But if $\rho_l^1 \neq \rho_l^2$ or $\rho_l^2 \neq \rho_l^3$ for some l, then the proof repeats that of sufficiency in Proposition 2.3.1 up to formula (3.4) (the latter formula holds for $j \neq l$), and then from part 2 to the end. Here the procedure described in part 2 leads to the equalities $'\varphi_j^1 = {'\varphi_j^2} = {'\varphi_j^3} = c_j$ for $j \neq l$ and $'\varphi_l^2 = {'\varphi_l^3}$, and in general $'\varphi_l^1 \neq {'\varphi_l^2}$ (in particular, it can happen that $'\varphi_l^1 < {'\varphi_l^2}$). Since the matrix E_l has the form

$$E_l = \begin{pmatrix} \rho_1^1 & 0 \quad 0 \\ 0 & \\ 0 & E_l' \end{pmatrix}$$

under the assumption that $\rho_l^1 \neq \rho_l^2$, the Fuchsian property of the system constructed again follows from Corollary 1.2.2. The case $\rho_l^2 \neq \rho_l^3$ is handled in exactly the same way.

PROPOSITION 2.3.2. *For any representation* (1.1.5) *of dimension* $p = 3$ *whose image lies in the set of upper triangular matrices the RH problem has a positive solution.*

PROOF. We consider a system (1.1.1) with regular singular points $a_1, \dots, a_n$ and the given monodromy (1.1.5) such that a fundamental matrix $T(y)$ of its solution space X has the form

$$T(y) = \begin{pmatrix} e_{11} & * & * \\ 0 & e_{22} & * \\ 0 & 0 & e_{33} \end{pmatrix}. \tag{3.8}$$

The existence of such a system was proved in part 2a) of Lemma 1.3.8. Denote by $c_i^j = \varphi_i(e_{jj})$ the normalization of the element $f_{jj}(y)$ at the point a_i. It can be assumed that

$$c_i^1 = c_i^2 = c_i^3 = 0, \qquad i = 1, \dots, \hat{l}, \dots, n, \tag{3.9}$$

where l is an arbitrary index in the set $\{1, \dots, n\}$. (Otherwise we apply the $\prod_{i\neq l} B(a_i, a_l, c_i^1, c_i^2, c_i^3)$ procedure to the space X and get a fundamental matrix of the form (3.8) with the condition (3.9).) The matrices G_i of monodromy operators g_i^* in the basis of columns of $T(y)$ have upper triangular form, and hence the same holds also for the matrices E_i in (1.1.9).

The rest of the proof reduces to an analysis of the following cases:

1) one of the matrices E_i is diagonalizable;

2) $\rho_i^1 = \rho_i^2$ for all i, or $\rho_i^2 = \rho_i^3$ for all i;

3) $\rho_i^1 = \rho_i^3$ for all i, and $\rho_l^1 \neq \rho_l^2$ for some l;

4) the remaining cases.

Case 1) was analyzed in Proposition 1.5.2. Case 2) follows from Corollary 2.3.1 and Proposition 2.2.1. Let us consider case 3). Note that in this case $\varphi_l(e_{11}) = \varphi_l(e_{33}) = c_l$ in view of (3.9) and the equality (1.2.22) for $e_{11}(y)$ and $e_{33}(y)$. We pass from the fundamental matrix $T(y)$ to $T_s(y) = T(y)S$, where S is an upper triangular matrix such that

$$S^{-1}E_lS = E_l^s = \begin{pmatrix} \rho_l^1 & 0 & 1 \\ 0 & \rho_l^2 & 0 \\ 0 & 0 & \rho_l^1 \end{pmatrix}. \tag{3.10}$$

Then

$$T_s(y) = U_l(z)(z-a_l)^{A_l}(y-a_l)^{E_l^s}, \tag{3.11}$$

where $A_l = \operatorname{diag}(c_l, \varphi_l(e_{22}), c_l)$, and $U_l(z)$ is an upper triangular meromorphic matrix with constant nonzero elements on the main diagonal.

Applying the

$$\prod_{i=1}^{n}(A(a_i, 2, 2, 0)\cdot A(a_i, 3, 3, 0))$$

procedure to X, we get a space $X' = \Gamma X$ having the following properties. In the decompositions (1.2.6) for the matrix $T_s'(y) = \Gamma T_s(y)$, $A_i = 0$ and $U_i(z)$ is a holomorphically invertible matrix for $i \neq l$. (By Corollary 1.2.2, X' is the solution space of a system (1.1.1) that is Fuchsian at $a_1, \dots, \hat{l}, \dots, a_n$.) For $i = l$ it follows from Lemma 1.3.2 that $U_l' = \Gamma(z)U_l$ is a holomorphically invertible matrix for the U_l in (3.11). Since the forms (3.11) and (3.10) of the matrices A_l and E_l^s tell us that

$$(z-a_l)^{A_l}E_l^s(z-a_l)^{-A_l}$$

is a holomorphic matrix, we again get from Corollary 1.2.2 that X' is the solution space of a system (1.1.1) that is Fuchsian at a_l.

We consider case 4). In this case there is a nonsingular upper triangular matrix S such that in the basis of columns of $T_s(y) = T(y)S$ the matrix E_l^s has one of the two forms

$$\text{4 a)}\ \begin{pmatrix} \rho_l^1 & 0 & 0 \\ 0 & \rho_l^2 & 1 \\ 0 & 0 & \rho_l^2 \end{pmatrix}, \qquad \text{4b)}\ \begin{pmatrix} \rho_1^l & 1 & 0 \\ 0 & \rho_l^1 & 0 \\ 0 & 0 & \rho_l^3 \end{pmatrix}.$$

Indeed, E_l^s must be nondiagonalizable (the case of diagonalizability is case 1)) and there must exist an l such that $\rho_l^1 \neq \rho_l^3$ (the condition that $\rho_l^1 = \rho_l^3$ for all l is case 3)), that is, either $\rho_l^1 \neq \rho_l^2 = \rho_l^3$ (case 4a)) or $\rho_l^1 = \rho_l^2 \neq \rho_l^3$ (case 4b)).

Let us consider case 4a). There is an index $k \neq l$ such that $\rho_k^1 = \rho_k^2 \neq \rho_k^3$ (otherwise we have case 2)). Consider the space $X|_{\tilde{O}_k}$. With the help of an upper triangular matrix S' we pass from the basis of columns of $T_s(y)$ to a Levelt basis with fundamental matrix $T_k(y)$. Then in the decompostion (1.2.6) for $T_k(y)$ we have that

$$E_k = \begin{pmatrix} \rho_k^1 & 1 & 0 \\ 0 & \rho_k^1 & 0 \\ 0 & 0 & \rho_k^3 \end{pmatrix}. \tag{3.12}$$

We remark that since S' is upper triangular, all the matrices in this decompostion again have upper triangular form; in particular, the matrices $T_s(y)$ and $T_k(y)$ have the form (3.2) a). According to Proposition 1.5.3 it can be assumed that the $\widetilde{T}(y)$ in (3.2) a) is now a fundamental matrix of a Fuchsian system (otherwise this can be achieved by an upper triangular transformation of the type (3.3) a); see Remark 1.5.1). Denote by $\mathbf{e}_2$ the second column of $T_s(y)$, and by J the set of indices for which $\varphi_i(e_{33}) > r$, where $r = \min(\varphi_i(e_{11}), \varphi_i(\mathbf{e}_2))$, $i \in J$.

We apply the $\prod_{i\in J} B(a_i, a_k, 0, 0, \varphi_i(e_{33} - r))$ procedure to the space X, and then the $\prod_{j=1}^{n} A(a_j, 3, 3, 0)$ procedure to the space X'. All the matrices are upper triangular in the decompositions (1.2.6) for the matrices $T_s'(y)|_{\widetilde{O}_i}$ with $i \neq k$ and for the matrix $T_k'|_{\widetilde{O}_k}$, and for the matrices A_i of normalizations with $i \neq k$

$$\varphi_i^3 \leq \min(\varphi_i^1, \varphi_i^2). \tag{3.13}$$

This and the fact that the fundamental matrix $\widetilde{T}(y)$ is Fuchsian (recall that $T'(y)$ has the form (3.2) a) and is also upper triangular) imply that all the conditions of Corollary 1.2.2 hold, and hence X' is the solution space of a system (1.1.1) that is Fuchsian at $a_1, \dots, \hat{k}, \dots, a_n$.

The condition (3.13) does not hold at a_k, but we again get from the form (3.12) of E_k that all the conditions in Corollary 1.2.2 hold. Hence, the system constructed is Fuchsian also at a_k.

Let us consider case 4b). There is an index $k \neq l$ such that $\rho_k^1 \neq \rho_k^2 = \rho_k^3$. The rest of the proof is a complete repetition of the proof of part 4a), with the indices k and l interchanged.

REMARK 2.3.1. It turns out that in larger dimensions the upper triangularity of the representation (1.1.5) does not ensure a positive solution for the RH problem. The first dimension known to the author in which this happens is $p = 7$, with $n = 4$ singular points (see Example 5.5.2).

PROPOSITION 2.3.3. *Suppose that a representation* (1.1.5) *of dimension* $p = 3$ *is such that one of the local representations* χ_l *in* (1.1.7) *decomposes into the direct sum of a one-dimensional and a two-dimensional representation. Then the RH problem has a positive solution for* (1.1.5).

PROOF. We consider a system (1.1.1) with regular singular points $a_1, \dots, a_n$ that is Fuchsian away from a_l and has the given monodromy (1.1.5). By (1.2.21) and (1.2.22), to prove the proposition it suffices to show that the sum Σ of the exponents of the solution space X of this system can always be increased.

We consider the decomposition (1.2.6) for the space $X|_{\widetilde{O}_l}$, after choosing as a Levelt basis a basis in which E_l has the form

$$E_l = \begin{pmatrix} \rho_l^1 & 0 & \varepsilon \\ 0 & \rho_l^2 & 1 \\ 0 & 0 & \rho_l^2 \end{pmatrix} \tag{3.14}$$

(The case when E_l is diagonalizable was analyzed in Proposition 1.5.2.) Here $\varepsilon = 0$ if $\rho_l^1 \neq \rho_l^2$, and ε can be nonzero for the case $\rho_l^1 = \rho_l^2$. Indeed, in the latter case an associated basis can be chosen as follows: first choose an associated basis for the subspace spanned by two linearly independent eigenvectors of the transformation g_l^*, and then complete it to form an associated basis in the whole of $X|_{\widetilde{O}_l}$. Here the number ε in (3.12) can be nonzero (see Remark 1.1.2, where an analogous situation is realized). By Lemma 1.3.9, it can be assumed that the matrix $U_l(z)$ now has the form (1.3.26) with i replaced by l. Suppose for definiteness that $c_{33} > 0$ in (1.3.26). The rest of the proof reduces to an analysis of the following cases:

1) $\varphi_l(u^l_{31}) \leq \varphi_l(u^l_{32}) < \infty$,
2) $\varphi_l(u^l_{31}) > \varphi_1(u^l_{32})$,
3) $u^l_{31}(z) \equiv u^l_{32}(z) \equiv 0$,

where u^l_{ij} are the elements of the matrix U_l in the decomposition (1.2.6).

By the $A(a_l, 3, 1, \varphi_l(u^l_{31}))$ procedure for X in case 1) and the $A(a_l, 3, 2, \varphi_l(u^l_{32}))$ procedure in case 2), we get a space X'. As follows from Remark 1.3.1, we get that $\Sigma' > \Sigma$ in both cases, where Σ' is the sum of the exponents of X'. (Here we have implicitly used the form (3.14) of E_l. If E_l were a Jordan block, then the second of the procedures described would not lead in the general case to an increase in the sum of the exponents, since for a Jordan basis the normalization of the second vector e_2 must not exceed the normalization of the first (Example 1.1.3); in the case (3.14) these normalizations are independent, since e_1 and e_2 are different eigenvectors of g^*_l.)

In case 3) it can be assumed that $u^l_{12}(a_l) = 0$ (this can always be achieved by replacing the first two rows of $U_l(z)$ by suitable linear combinations of these rows, which corresponds to passing from the space X to $X' = SX$, where

$$S = \begin{pmatrix} \multicolumn{2}{c}{\widetilde{S}} & \begin{matrix} 0 \\ 0 \end{matrix} \\ 0 & 0 & 1 \end{pmatrix}$$

is a nonsingular constant matrix).

Case 3) can be analyzed in two subcases:

3a) $u^l_{12}(z) \not\equiv 0$;
3b) $u^l_{12}(z) \equiv 0$.

Applying the $A(a_l, 3, 3, c_{33})$ procedure and the $A(a_l, 1, 2, \varphi_l(u^l_{12}))$ procedure to X in case 3a), we again get from Remark 1.3.1 and the form (3.14) of E_l that $\Sigma' > \Sigma$.

In case 3b) we get from the form (3.14) of E_l that the elements $e_{12}(y)$, $e_{31}(y)$, and $e_{32}(y)$ of $T_l(y)$ are identically equal to zero; that is, after a transposition of the first two columns (transition to another basis in X) and of the first two rows (a gauge transformation of X) the matrix $T_l(y)$ assumes the upper triangular form $T'_l(y)$. Hence, the image of the homomorphism (1.1.5) lies in the set of upper triangular matrices. Proposition 2.3.3 follows in this case from Proposition 2.3.2.

We get the following theorem from Propositions 2.3.1–2.3.3 and Theorem 2.1.1.

THEOREM 2.3.1. *The RH problem for a representation* (1.1.5) *of dimension* $p = 3$ *has a negative solution if and only if the following three conditions hold*:

a) *each of the generators* G_i *of the monodromy group* $\operatorname{Im}\chi$ *reduces to a Jordan block*;

b) *the representation* (1.1.5) *is reducible*;

c) *the Fuchsian weight* γ_{χ^2} *of the corresponding two-dimensional subrepresentation* (*or factor representation*) *is greater than zero.*

REMARK 2.3.2. It follows from conditions a), b), and c) that the two-dimensional subrepresentation (factor representation) χ^2 is irreducible, for otherwise by a) it is commutative, and the generators $\tilde{G}_i$ of the group $\operatorname{Im}\chi^2$ have the form indicated in the proof of Proposition 2.2.1. From the proof of this proposition we get that $\gamma_{\chi^2} = 0$, which contradicts the condition $\gamma_{\chi^2} > 0$. Hence, χ^2 is irreducible.

THEOREM 2.3.2. *For any three points a_1, a_2, and a_3 and any representation* (1.1.5) *of dimension $p = 3$ the RH problem has a positive solution.*

PROOF. By Theorem 2.3.1 and Remark 2.3.2, it suffices to consider a reducible representation (1.1.5) of dimension $p = 3$ for which all the matrices G_i reduce to Jordan blocks, and the corresponding two-dimensional subrepresentation or factor representation χ^2 is irreducible. It follows from the reducibility of (1.1.5) that for the corresponding one-dimensional subrepresentation or factor representation the number $r = \sum_{i=1}^{3} \rho_i$ is an integer, where ρ_i is the eigenvalue of the Jordan block E_i. Therefore, we conclude from Proposition 2.2.1 that γ_{χ^2} is even. It follows from the irreducibility of χ^2 and Proposition 2.2.3 for $n = 3$ that $\gamma_{\chi^2} \leq 1$; therefore, $\gamma_{\chi^2} = 0$. Theorem 2.3.2 now follows from Theorem 2.3.1.

THEOREM 2.3.3. *For any $n > 3$, any set of points $a_1, \dots, a_n$, and any $p \geq 3$ there exists a representation* (1.1.5) *for which the RH problem has a negative solution.*

PROOF. 1. We prove the theorem first for the case $p = 3$. To do this it suffices to construct a representation (1.1.5) satisfying the conditions of Theorem 2.3.1.

We consider the system (1.1.1), (2.1). Since the number of points is greater than three, there are 2-vectors $b_2, \dots, b_n$ such that $b_j \neq 0$, $j = 2, \dots, n$, and

$$\sum_{j=2}^{n} b_j = 0, \qquad \operatorname{rank} {}'B^j = 2, \qquad j = 2, \dots, n, \tag{3.15}$$

where the ${}'B^i$ denote the matrices

$${}'B^1 = \begin{pmatrix} 0 & z^{-\gamma'} & 0 \\ 0 & \multicolumn{2}{c}{\multirow{2}{*}{B^1}} \\ 0 & & \end{pmatrix}, \qquad {}'B^i = \begin{pmatrix} 0 & b_i \\ 0 & \multirow{2}{*}{B^i} \\ 0 & \end{pmatrix}, \qquad i > 1, \tag{3.16}$$

the B^i are the matrices in (2.39), $\gamma' = \frac{1}{2}\gamma_\chi$, and γ_χ is the Fuchsian weight of the monodromy representation (1.1.5) of the system (1.1.1), (2.1), $\gamma_\chi > 0$.

Consider the system (1.1.1), (2.1), (3.16). By (3.15), it satisfies the condition (2.2). We show that the monodromy representation of this system satisfies the conditions of Theorem 2.3.1.

It follows from the construction of the system that conditions b) and c) of Theorem 2.3.1 are satisfied for it.

Since the matrices ${}'B^i$ in (3.16) reduce to Jordan blocks for $i > 1$, we get from (1.2.3) and (1.2.11) that the same is true for the matrices E_i, and hence for $G_i = \exp(2\pi i E_i)$.

We consider the solution space of the system (1.1.1), (2.1), (3.16) in $\tilde{O}_1$. Let $T_1(y)$ be the fundamental matrix in (2.41) for the system (1.1.1), (2.1), (2.39). Then the matrix

$$T = \begin{pmatrix} 1 & e_{12} & e_{13} \\ 0 & \multicolumn{2}{c}{\multirow{2}{*}{T_1}} \\ 0 & & \end{pmatrix} \tag{3.17}$$

is the fundamental matrix for the system (1.1.1), (2.1), (3.16). Substituting $T(y)$ in this system, we get

$$(3.18)\qquad \begin{cases} \dfrac{de_{12}}{dz} = z^{-\gamma'-1}e_{22} + \displaystyle\sum_{i=2}^{n}\frac{1}{z-a_i}(b^i_{12}e_{22} + b^i_{13}e_{32}), \\ \dfrac{de_{13}}{dz} = z^{-\gamma'-1}e_{23} + \displaystyle\sum_{i=2}^{n}\frac{1}{z-a_i}(b^i_{11}e_{23} + b^i_{13}e_{33}). \end{cases}$$

Since it follows from Remark 2.2.2 that

$$\varphi_1(e_{22}) = \gamma', \qquad \varphi_1(e_{23}) \geq 0, \qquad \varphi_1(e_{33}) = -\gamma', \qquad \varphi_1(e_{32}) \geq \gamma',$$

where $\varphi_1(f)$ denotes the normalization of f at the point $a_1 = 0$, we get from (3.18) that

$$(3.19)\qquad \varphi_1\left(\frac{de_{12}}{dz}\right) = -1, \qquad \varphi_1\left(\frac{de_{13}}{dz}\right) = -\gamma' - 1.$$

It follows from (3.19) that $a_1 = 0$ is a regular singular point for the system (1.1.1), (2.1), (3.16).

The matrix E_1 in the decomposition (1.2.6) for $T(y)$ has the form

$$(3.20)\qquad E_1 = \begin{pmatrix} 0 & \alpha & \beta \\ 0 & 0 & 1 \\ 0 & 0 & 0 \end{pmatrix}.$$

If α were zero, then the matrix y^{E_1} would have the form

$$y^{E_1} = \begin{pmatrix} 1 & 0 & \beta \ln y \\ 0 & 1 & \ln y \\ 0 & 0 & 1 \end{pmatrix},$$

and the element e_{12} in (3.17) would be a single-valued meromorphic function in a neighborhood of $a_1 = 0$; but then the number $\varphi_1(de_{12}/dz)$ would be nonnegative in the case when $e_{12}(z)$ is holomorphic, and it would be less than -1 in the case when e_{12} has a pole at $a_1 = 0$. We get a contradiction to the first equality in (3.19). Hence, $\alpha \neq 0$ in (3.20), and E_1 reduces to a Jordan block. Therefore, the same is true also for G_1. The condition a) of Theorem 2.3.1 is satisfied.

Thus, the RH problem has a negative solution for the monodromy representation (1.1.5) of the system (1.1.1), (2.1), (3.16).

2. We consider 3-vectors $'b_1 = (1, 0, 0), 'b_2, \dots, 'b_n$ such that

$$\sum_{j=1}^{n} b_j = 0, \qquad \text{rank}'' B^j = 2, \qquad j = 2, \dots, n,$$

where the $''B^i$ denote the matrices

$$''B^i = \begin{pmatrix} 0 & b_i \\ 0 & \\ 0 & 'B^i \\ 0 & \end{pmatrix},$$

and the $'B^i$ denote the matrices in (3.16).

Denote by J_0 the $(p-4)\times(p-4)$ Jordan block with eigenvalue 0, by J_q the matrix $J_q = qJ_0$, and by R_q the $(p-4)\times 4$ matrix with unique nonzero element $r^q_{p-4,1} = q$. We consider the matrices

$$(3.21)\qquad {}'_pB^1 = \begin{pmatrix} J_{1-n} & R_{1-n} \\ 0 & {}''B^1 \end{pmatrix}, \qquad {}'_pB^i = \begin{pmatrix} J_1 & R_1 \\ 0 & {}''B^i \end{pmatrix}, \qquad i = 2, \dots, n.$$

Just as in part 1, it can be proved that the monodromy matrices ${}_pG_i$ of the system (1.1.1), (2.1), (3.21) reduce to Jordan blocks for $k > 1$. Since the matrix formed by the intersections of the first $p-3$ rows and the columns of the matrix ${}'_pB_1$ is a Jordan block (multiplied by $(1-n)$), it follows from (1.2.3) and (1.2.11) that

$${}_pE_1 = \begin{pmatrix} J_{1-n} & R_{1-n} \\ 0 & E_1 \end{pmatrix},$$

but by part 1 the matrix E_1 is upper triangular and reduces to a Jordan block. Consequently, ${}_pE_1$ (and ${}_pG_1$) also reduces to a Jordan block.

If the representation (4) with monodromy matrix ${}_pG_i$ could be realized as the monodromy representation of some Fuchsian system, then by Corollary 1.4.1 the sum of the exponents of the $(p-2)$-dimensional subspace X_{p-2} of solutions of this system would be zero. But in this case Theorem 5.1.2 (which will be proved in Chapter V, but whose proof is based only on the facts above) would give us that by a substitution (1.3.1) with a constant matrix $\Gamma = S$ this system could be reduced to a Fuchsian system with factor system (1.1.1), (2.1), (2.39) having equal normalizations at all points, which is impossible. Therefore, for the monodromy representation (1.1.5) of the system (1.1.1), (2.1), (3.21) there is no Fuchsian system realizing it. The theorem is proved.

In view of (1.2.9) we get the following result from Corollary 2.2.1 and the proof of Theorem 2.3.3.

COROLLARY 2.3.2. *Suppose that a representation* (1.1.5) *of dimension* $p = 3$ *satisfies conditions* a), b), *and* c) *of Theorem* 2.3.1. *Then there exists a system* (1.1.1) *with monodromy* (1.1.5) *and with regular singular points* $a_1, \dots, a_n$ *that is Fuchsian at all points except one, where the order of the pole of the coefficient matrix is equal to* $\gamma_{\chi^2}/2 + 1$, *and there is no analogous system with pole of lower order.*

Proposition 2.2.3 and Corollary 2.3.2 give us

COROLLARY 2.3.3. *For any representation* (1.1.5) *of dimension* $p = 3$ *there exists a system* (1.1.1) *with regular singular points* $a_1, \dots, a_n$ *and monodromy* (1.1.5) *that is Fuchsian at all the points except perhaps one, where the order of the pole of the coefficient matrix does not exceed* $[n/2]$ ($[x]$ *is the integer part of a number* x).

4. Example of a negative solution of the Riemann–Hilbert problem

In this section we analyze a concrete example of a representation (1.1.5) for which the RH problem has a negative solution. Here we make as few references as possible to the theorems of general character proved above. In fact, in the proofs of the assertions of this section the corresponding theorems and propositions will be reproved in simplified versions adapted to the concrete situation at hand.

We consider the system

$$
\begin{aligned}
(4.1)\qquad df = \Bigg(& \begin{pmatrix} 0 & 1 & 0 \\ 0 & z & 0 \\ 0 & 0 & -z \end{pmatrix} \frac{dz}{z^2} + \frac{1}{6} \begin{pmatrix} 0 & 6 & 0 \\ 0 & -1 & 1 \\ 0 & -1 & 1 \end{pmatrix} \frac{dz}{z+1} \\
& + \frac{1}{2} \begin{pmatrix} 0 & 0 & 2 \\ 0 & -1 & -1 \\ 0 & 1 & 1 \end{pmatrix} \frac{dz}{z-1} + \frac{1}{3} \begin{pmatrix} 0 & -3 & -3 \\ 0 & -1 & 1 \\ 0 & -1 & 1 \end{pmatrix} \frac{dz}{z-1/2} \Bigg) f.
\end{aligned}
$$

This system is Fuchsian at the points -1, 1, $1/2$, and the point ∞ is a point of holomorphy for it, since the residue at infinity of the form on the right-hand side of (4.1) is 0. It will be proved below that 0 is a regular singular point for (4.1).

A fundamental matrix $T(y)$ of the system (4.1) can be written in the form

$$
(4.2)\qquad T(y) \doteq \begin{pmatrix} 1 & u(y) \quad v(y) \\ 0 & \\ 0 & T'(y) \end{pmatrix},
$$

where $T'(y) = ((t_{ij}))$ is a fundamental matrix of the Fuchsian system

$$
\begin{aligned}
(4.3)\qquad df = \Bigg(& \begin{pmatrix} 1 & 0 \\ 0 & -1 \end{pmatrix} \frac{dz}{z} + \frac{1}{6} \begin{pmatrix} -1 & 1 \\ -1 & 1 \end{pmatrix} \frac{dz}{z+1} \\
& + \frac{1}{2} \begin{pmatrix} -1 & -1 \\ 1 & 1 \end{pmatrix} \frac{dz}{z-1} + \frac{1}{3} \begin{pmatrix} -1 & 1 \\ -1 & 1 \end{pmatrix} \frac{dz}{z-1/2} \Bigg) f
\end{aligned}
$$

with singular points 0, -1, 1, $1/2$.

The monodromy representation

$$
(4.4)\qquad \chi^* : \pi_1(CP^1 \setminus \{0, -1, 1, 1/2\}, z_0) \to GL(3; C)
$$

of the system (4.1) is an example of a representation (1.1.5) for which the RH problem has a negative solution. In other words, we have

THEOREM 2.4.1. *There does not exist a Fuchsian system with singular points* 0, -1, 1, $1/2$ *whose monodromy coincides with that of the system* (4.1).

The idea of the proof consists in the following. The monodromy of (4.1) is reducible (the vector $(1, 0, 0)^T$ is a solution of the system (4.1) and generates a one-dimensional subspace that is invariant under the action of the monodromy operators). Denote by χ^2 the two-dimensional factor representation of the representation χ^*; χ^2 is the monodromy representation of the system (4.3) with fundamental matrix $T'(y)$ in (4.2).

As will be shown below, a positive solution of the RH problem for the representation χ^* yields a positive solution of the RH problem for the representation χ^2 with some additional restrictions, but it turns out that this problem does not have a solution for χ^2. The original problem (the RH problem for the representation χ^*) thus has a negative solution.

We prove the theorem in a series of lemmas.

Let us first study the system (4.3).

LEMMA 2.4.1. *Each of the monodromy matrices* G' *of the representation* χ^2 *reduces to a Jordan block. The Fuchsian weight* γ_{χ^2} *of* χ^2 *is equal to* 2.

PROOF. Since the exponents β_i^j of the system (4.3) coincide with the eigenvalues of the coefficient matrices of this system (Corollary 1.2.3), it follows from the definition (1.2.5) of the exponents that

$$\begin{gathered}\rho_i^1 = \rho_i^2 = 0, \qquad i = 1, 2, 3, 4, \\ \varphi_1^1 = 1, \qquad \varphi_1^2 = -1, \qquad \varphi_i^1 = \varphi_i^2 = 0, \quad i = 2, 3, 4,\end{gathered} \tag{4.5}$$

where φ_1^1 and φ_1^2 denote the normalizations of the solution spaces X at 0.

It follows immediately from (4.5) that the Fuchsian weight γ_ω of the system (4.3) is equal to 2.

We consider the monodromy matrices G_2', G_3', and G_4' corresponding to a circuit of the points -1, 1, $1/2$. Since all the eigenvalues of the matrices $'B^2 = \begin{pmatrix} -1 & 1 \\ -1 & 1 \end{pmatrix} = {}'B^4$ and $'B^3 = \begin{pmatrix} -1 & -1 \\ 1 & 1 \end{pmatrix}$ are equal to zero, no two of them differ by a positive integer (the nonresonance case), and hence the matrices G_2 and G_4 are similar to the matrix $\exp(2\pi i\, 'B^2)$, and G_3 is similar to the matrix $\exp(2\pi i\, 'B^3)$ (see [16]). Since each of the matrices $'B^2$ and $'B^3$ reduces to a Jordan block, each of the monodromy matrices G_2', G_3', and G_4' reduces to a Jordan block.

We consider the decomposition (1.2.6) for the fundamental matrix $T'(y)$ of the solution space $X|_{\widetilde{O}_1}$ of the system (4.3):

$$T'(y) = U_1(z) z^{\begin{pmatrix} 1 & 0 \\ 0 & -1 \end{pmatrix}} y^{E_1}, \tag{4.6}$$

where $E_1 = \begin{pmatrix} 0 & \varepsilon \\ 0 & 0 \end{pmatrix}$.

From (4.6) and (4.3) we obtain

$$\begin{aligned}\begin{pmatrix} 1 & 0 \\ 0 & -1 \end{pmatrix} &= \lim_{z\to 0} z \frac{dT'(y)}{dz} (T'(y))^{-1} \\ &= \lim_{z \to 0} \left(z \frac{dU_1(z)}{dz} U_1^{-1}(z) + U_1(z) \begin{pmatrix} 1 & \varepsilon z^2 \\ 0 & -1 \end{pmatrix} U_1^{-1}(z) \right) \\ &= U_1(0) \begin{pmatrix} 1 & 0 \\ 0 & -1 \end{pmatrix} U_1^{-1}(0).\end{aligned} \tag{4.7}$$

It follows from (4.7) that

$$U_1(0) = \begin{pmatrix} c_1 & 0 \\ 0 & c_2 \end{pmatrix}.$$

We pass from the matrix $T'(y)$ to

$$\widetilde{T}' = T'(y) \begin{pmatrix} 1/c_1 & 0 \\ 0 & 1/c_2 \end{pmatrix},$$

and denote it again by $T'(y)$. For this new matrix the matrices U_1 and E_1 in the decomposition (4.6) have the form

$$U_1(z) = \begin{pmatrix} 1 & c z^m \\ s z^k & 1 \end{pmatrix} (1 + o(1)), \qquad m, k > 0, \tag{4.8}$$

$$E_1 = \begin{pmatrix} 0 & \alpha \\ 0 & 0 \end{pmatrix}, \qquad \alpha = \frac{c_1}{c_2} \varepsilon. \tag{4.9}$$

In a neighborhood of 0 the right-hand side ωf of the system (4.3) has by (4.6), (4.8), and (4.9) the form

$$\begin{aligned}(4.10)\quad \omega = dT'(T')^{-1} &= \left(\frac{dU_1(z)}{dz}U_1^{-1}(z) + \frac{U_1(z)}{z}\begin{pmatrix}1 & \alpha z^2\\ 0 & -1\end{pmatrix}U_1^{-1}(z)\right)dz\\ &= \left(\begin{pmatrix}* & mcz^{m-1}\\ * & *\end{pmatrix} + \begin{pmatrix}* & \alpha z - 2cz^{m-1}\\ * & *\end{pmatrix}\right)(1+o(1))\,dz,\end{aligned}$$

since

$$U_1^{-1}(z) = \begin{pmatrix}1 & -cz^m\\ -sz^k & 1\end{pmatrix}(1+o(1)).$$

On the other hand, we have from the explicit form of the right-hand side ωf of (4.3) that

(4.11)

$$\begin{aligned}\omega &\left[\begin{pmatrix}1 & 0\\ 0 & -1\end{pmatrix}\frac{1}{z} + \frac{1}{6}\begin{pmatrix}-1 & 1\\ -1 & 1\end{pmatrix}(1 - z + o(z))\right.\\ &\left.+\frac{1}{2}\begin{pmatrix}-1 & -1\\ 1 & 1\end{pmatrix}(1 + z + o(z)) + \frac{1}{3}\begin{pmatrix}-1 & 1\\ -1 & 1\end{pmatrix}(-2)(1 + 2z + o(z))\right]dz\\ &= \begin{pmatrix}* & -z\\ * & *\end{pmatrix}(1+o(1))\,dz.\end{aligned}$$

According to (4.10) and (4.11),

$$[c(m-2)z^{m-1} + \alpha z](1+o(1)) = -z(1+o(1)),$$

from which we get

$$(4.12)\qquad m \geq 2$$

and $\alpha = -1$.

From the equality $\alpha = -1$ we get that ${}'G_1 = \exp(2\pi i E_1)$ is a Jordan block. The first part of the lemma is proved.

We show that the Fuchsian weight of χ^2 is equal to 2. Assume the contrary. Suppose that there exists a Fuchsian system

$$(4.13)\qquad df = \widetilde{\omega} f$$

with the same monodromy χ^2 whose Fuchsian weight $\gamma_{\widetilde{\omega}}$ is zero (the parity of the Fuchsian weight is preserved (see Proposition 2.2.1); therefore, it follows from the inequality $\gamma_{\widetilde{\omega}} < 2$ that $\gamma_{\widetilde{\omega}} = 0$). The latter means that the normalizations of this system at each point are equal:

$$(4.14)\qquad \widetilde{\varphi}_i^1 = \widetilde{\varphi}_i^2, \qquad i = 1, 2, 3, 4.$$

In the system (4.13) we make the change of unknown function $g = \Gamma f$,

$$(4.15)\qquad \Gamma(z) = \frac{(z+1)^{-\widetilde{\varphi}_2^1}(z-1)^{-\widetilde{\varphi}_3^1}(z-1/2)^{-\widetilde{\varphi}_4^1}}{z^{-\Sigma}}I,$$

where $\Sigma = \sum_{i=2}^4 \widetilde{\varphi}_i^1$, and obtain the system $dg = \check{\omega} g$.

Since $2\Sigma + 2\widetilde{\varphi}_1^2 = 0$ by the residue theorem for the form $\operatorname{tr}\widetilde{\omega}$, it follows that $-\Sigma = \widetilde{\varphi}_1^1$, and hence the normalizations of the system $dg = \check{\omega} g$ are equal to zero at all the points.

We consider the matrix $T'(y)$ in (4.6) and the fundamental matrix $\check{T}(y)$ of the system $dg = \check{\omega} g$ with the same monodromy; that is,

$$\check{T}^{-1}(y)(g_i^*\check{T})(y) = (T')^{-1}(g_i^* T')(y) = G_i', \qquad i = 1, \dots, 4. \tag{4.16}$$

The decompositions (1.2.6) for $\check{T}(y)$ at the points 0, -1, 1, $1/2$ have the form

$$\check{T}(y) = \check{U}_i(z) y^{E_i}, \qquad \det \check{U}_i(a_i) \neq 0.$$

The matrix $T'(y)$ can be represented in the same form at the points -1, 1, $1/2$. Therefore, the single-valued (in view of (4.16)) meromorphic matrix-valued function $Z(z) = T'(y)\check{T}^{-1}(y)$ is holomorphically invertible on CP^1 away from the point $a_1 = 0$. By (4.8), $Z(z)$ has in a neighborhood of $a_1 = 0$ the form

$$\begin{aligned} Z(z) &= \begin{pmatrix} 1 & cz^m \\ sz^k & 1 \end{pmatrix} z^{\left(\begin{smallmatrix} 1 & 0 \\ 0 & -1 \end{smallmatrix}\right)} \check{U}^{-1}(z)(1 + o(1)) \\ &= \begin{pmatrix} z & cz^{m-1} \\ * & * \end{pmatrix} \begin{pmatrix} u_{11}(z) & u_{12}(z) \\ u_{21}(z) & u_{22}(z) \end{pmatrix} (1 + o(1)). \end{aligned}$$

From this, by the condition (4.12), we get for the first row $\mathbf{z}_1$ of $Z(z)$ that

$$\mathbf{z}_1(z) = z\mathbf{t}(z),$$

where $\mathbf{t}(z)$ is a holomorphic row vector at $a_1 = 0$. We conclude from Liouville's theorem that $\mathbf{z}_1(z) \equiv 0$, which contradicts the holomorphic invertibility of $Z(z)$ away from $a_1 = 0$. This contradiction means that the Fuchsian weight of the representation χ^2 is 2.

REMARK 2.4.1. From (4.6), (4.8), and (4.12) we get for the elements t_{ij} of $T'(y)$ that

$$\varphi_1(t_{11}) = 1, \qquad \varphi_1(t_{12}) \geq 1, \qquad \varphi_1(t_{21}) \geq 2, \qquad \varphi_1(t_{22}) = -1. \tag{4.17}$$

Let us return to the representation χ^* and the system (4.1).

LEMMA 2.4.2. *The point* 0 *is a regular singular point for the system* (4.1). *Each of the monodromy matrices* G_i *of the system* (4.1) *reduces to a Jordan block.*

PROOF. To prove the first assertion of the lemma it suffices to show that the elements $u(y)$ and $v(y)$ of the matrix $T(y)$ in (4.2) have finite order at zero. Substituting (4.2) in (4.1), we get

$$\begin{aligned} \frac{du}{dz} &= t_{11}(y)\left(\frac{1}{z^2} + \frac{1}{z+1} - \frac{1}{z-1/2}\right) + t_{21}(y)\left(\frac{1}{z-1} - \frac{1}{z-1/2}\right), \\ \frac{dv}{dz} &= t_{12}(y)\left(\frac{1}{z^2} + \frac{1}{z+1} - \frac{1}{z-1/2}\right) - t_{22}(y)\left(\frac{1}{z-1} - \frac{1}{z-1/2}\right). \end{aligned} \tag{4.18}$$

From this and (4.17),

$$\varphi_1\left(\frac{du}{dz}\right) = \varphi_1\left(t_{11}\frac{1}{z^2}\right) = -1, \qquad \varphi_1\left(\frac{dv}{dz}\right) \geq -1, \tag{4.19}$$

and hence $u(y)$ and $v(y)$ do have finite order at zero.

We proceed to the proof of the second part of the lemma. The fact that the matrices G_2, G_3, and G_4 reduce to Jordan blocks can be proved exactly like the analogous assertion for the matrices G_2', G_3', and G_4' in Lemma 2.4.1.

The matrix E_1 in the decomposition (1.2.6) for the matrix $T(y)$ in (4.2) has the form

$$E_1 = \begin{pmatrix} 0 & \alpha & \beta \\ 0 & 0 & \varepsilon \\ 0 & 0 & 0 \end{pmatrix}, \qquad \varepsilon \neq 0. \tag{4.20}$$

By (4.6) and (4.2),

$$T(y) = U(z) z^{A_1} y^{E_1},$$

where $U(z)$ is holomorphic (but not holomorphically invertible) at 0, A_1 is the integer matrix of normalizations at 0 of the columns of $T(y)$, and E_1 is from (4.20). If $\alpha = 0$, then

$$y^{E_1} = \begin{pmatrix} 1 & 0 & \beta \ln y \\ 0 & 1 & \varepsilon \ln y \\ 0 & 0 & 1 \end{pmatrix},$$

and the element $u(y)$ in (4.2) of $T(y)$ is a single-valued function in a neighborhood of 0. Therefore, the normalization $\varphi_1(du/dz)$ is either nonnegative (if u is a holomorphic function), or does not exceed -2 (if $\varphi_1(u) \leq -1$). However, it follows from (4.19) that $\varphi_1(du/dz) = -1$. This contradiction means that $\alpha \neq 0$ in (4.20); that is, E_1 and G_1 reduce to a Jordan block. The lemma is proved.

PROOF OF THEOREM 2.4.1. Assume the contrary. Suppose that there exists a Fuchsian system

$$df = \omega f \tag{4.21}$$

with monodromy (4.4). In the solution space X of this system we consider a common eigenvector t_1 for the monodromy matrices $G_1, \dots, G_4$. Since all the eigenvalues of the monodromy operators are equal to 1, t_1 is single-valued meromorphic function on CP^1. By the residue theorem for each component of the form $(dt_{11}/t_{11}, dt_{12}/t_{12}, dt_{13}/t_{13})^T$,

$$\sum_{i=1}^{4} \varphi_i(t_{1j}) = 0, \qquad j = 1, \dots, 4,$$

and hence

$$\sum_{i=1}^{4} \varphi_i(t_1) \leq 0. \tag{4.22}$$

Since each matrix G_i reduces to a Jordan block, an associated basis of X in a neighborhood of a_i is a Jordan basis (see Example 1.1.3), and the first element of such a basis for all i is the vector-valued function $t_1(z)$. Denote the associated basis at a_i by $t_1(z)$, $t_2^i(y)$, $t_3^i(y)$. Then (see Example 1.1.3)

$$\varphi_i(t_1) \geq \varphi_i(t_2^i) \geq \varphi_i(t_3^i). \tag{4.23}$$

Since the eigenvalues of the coefficient matrices $B^i = \operatorname{res}_{a_i} \omega$ of the system (4.21) coincide with the exponents β_i^j of the space X of this system, and since those in turn are equal to the normalizations φ_i^j (because $\rho_1 = \cdots = \rho_4 = 0$), it follows from the residue theorem for the form $\operatorname{tr} \omega$ that

$$\sum_{i=1}^{4} (\varphi_i(t_1) + \varphi_i(t_2^i) + \varphi_i(t_3^i)) = 0 \tag{4.24}$$

(see also Proposition 1.2.3). From (4.22), (4.23), and (4.24) we conclude immediately that

$$\varphi_i(t_1) = \varphi_i(t_2^i) = \varphi_i(t_3^i), \qquad i = 1, \dots, 4, \tag{4.25}$$

$$\sum_{i=1}^{4} \varphi_i(t_1) = 0. \tag{4.26}$$

We consider a nonsingular constant matrix S such that

$$t_1'(z)|_{z=0} = S \cdot t_1(z)|_{z=0} = (c, 0, 0)^T. \tag{4.27}$$

In this case it follows from (4.26) that $t_1'(z) = (t_{11}'(z), 0, 0)^T$. Indeed, from (4.26),

$$\varphi_i(t_{11}'(z)) \leq \varphi_i(t_{1j}'(z)), \qquad j = 2, 3, \ i = 1, \dots, 4, \tag{4.28}$$

for if not $\varphi_l(t_1'(z)) < \varphi_l(t_{11}'(z))$ for some l, and $\sum_{i=1}^4 \varphi_i(t_1'(z)) < \sum_{i=1}^4 \varphi_i(t_{11}'(z)) = 0$, which contradicts (4.26). But we get from (4.28) and the condition (4.27) that $0 = \varphi_1(t_{11}') < \varphi_1(t_{1j}')$ and $0 = \sum_{i=1}^4 \varphi_i(t_1') < \sum_{i=1}^4 \varphi_i(t_{1j}')$, $j = 2, 3$; that is, the degree of the divisor of each of the functions $t_{1j}'(z)$, $j = 2, 3$, is strictly greater than zero; hence, $t_{12}'(z) = t_{13}'(z) \equiv 0$.

In the system (4.21) we make the substitution $g = Sf$ and denote the form of the resulting system again by ω. The latter system now has a fundamental matrix of the form (4.2). By (4.25), the normalizations of the columns of $T'(y)$ in (4.2) are equal at all the points. The latter shows that the Fuchsian weight $\gamma_{\omega'}$ of the system (4.21) with fundamental matrix $T'(y)$ is equal to 0, and this in turn means that the Fuchsian weight of the two-dimensional factor representation χ^2 of χ^* is equal to 0, which contradicts the condition $\gamma_{\chi^2} = 2$ proved in Lemma 2.4.1. It follows from this contradiction that there is no Fuchsian system with singular points 0, -1, 1, $1/2$ whose monodromy coincides with that of the system (4.1). The theorem is proved.

Another variant of the proof of Theorem 2.4.1 is given in [28].

5. The Fuchsian weight of a representation and isomonodromy deformations

Denote by $\mathscr{F}_n$ a group with n generators $h_1, \dots, h_n$ and with unique relation $h_1 \cdot \dots \cdot h_n = e$, and by $\varkappa_a$ the isomorphism

$$\varkappa_a : \pi_1(CP^1 \setminus \mathscr{D}, z_0) \to \mathscr{F}_n \tag{5.1}$$

with $\varkappa_a(g_i) = h_i$ for the generators g_i defined in §1 of Chapter I.

Any representation (1.1.5) can be written in the form $\chi = \chi' \circ \varkappa_a$, where

$$\chi' : \mathscr{F}_n \to GL(p; C) \tag{5.2}$$

is some representation of $\mathscr{F}_n$. In what follows we denote χ by $\chi(a)$, where $a = (a_1, \dots, a_n)$.

We consider the dependence of the Fuchsian weight $\gamma_{\chi(a)}$ on the choice of the points a.

The following proposition is obvious.

PROPOSITION 2.5.1. *If for the tuples* $a = (a_1, \dots, a_n)$ *and* $b = (b_1, \dots, b_n)$ *there is a conformal transformation* $r: CP^1 \to CP^1$ *with* $r(a_i) = r(b_i)$ *and* $r(\gamma_i^a) = r(\gamma_i^b)$, $i = 1, \dots, n$, *for the paths* γ_i *defined in §1 of Chapter* I, *then*

$$\gamma_{\chi(a)} = \gamma_{\chi(b)}.$$

PROPOSITION 2.5.2. *Suppose that the Fuchsian weight* $\gamma_{\chi(a)}$ *of a representation* $\chi(a)$ *with nondiagonalizable monodromy matrices* $\chi'(h_1), \dots, \chi'(h_n)$ *is greater than* 1. *There exist an* $\varepsilon > 0$ *and an index* l *such that the inequality* $\gamma_{\chi(a')} < \gamma_{\chi(a)} - 1$ *holds for any tuple* $a' = (a_1, \dots, a_{l-1}, a_l + t, a_{l+1}, \dots, a_n)$ *of points with* $0 < |t| < \varepsilon$.

PROOF. Let us consider a Fuchsian system (1.1.1) with monodromy (1.1.5) and Fuchsian weight $\gamma = \gamma_{\chi(a)}$ that is reduced to the form (2.11)–(2.15) at some point a_l. It follows from Proposition 2.2.2 that in (2.15) either $c \neq 0$ and $m \geq \gamma$, or $c = 0$. We use a conformal transformation to carry the point a_l to 0 and the point ∞ to ∞, and we denote the resulting tuple again by $a_1, \dots, a_n$.

Consider the isomonodromy deformation

$$df = \left[\left(\frac{B^1(t)}{z-t} + \sum_{i=2}^{n} \frac{B^i(t)}{z-a_i}\right) dz\right] f \tag{5.3}$$

of the constructed system for small values of t, where

$$\frac{dB^i(t)}{dt} = \frac{1}{a_i - t}[B^i(t), B^1(t)], \qquad B^i(0) = B^i, \qquad \sum_{i=1}^{n} B^i(t) = 0. \tag{5.4}$$

(Equations (5.4) are the equations of the isomonodromy deformation of Schlesinger [56].)

From (2.14), (2.15), and (1.2.11),

$$B^1(0) = \begin{pmatrix} b+\gamma_1 & 0 \\ 0 & b-\gamma_2 \end{pmatrix},$$

and thus (5.4) gives us that

$$\frac{dB^i(0)}{dt} = \frac{1}{a_i}\begin{pmatrix} 0 & -\gamma b^i_{12} \\ \gamma b^i_{21} & 0 \end{pmatrix}$$

and

$$B^i(t) = \begin{pmatrix} b^i_{11} + o(t) & b^i_{12} - \frac{\gamma}{a_i} b^i_{12} t + o(t) \\ b^i_{21} + \frac{\gamma}{a_i} b^i_{21} t + o(t) & b^i_{22} + o(t) \end{pmatrix}. \tag{5.5}$$

From this and again from (5.4), we get

$$B^1(t) = -\sum_{i=2}^{n} B^i(t) = \begin{pmatrix} b+\gamma_1 + o(t) & o(t) \\ o(1) & b-\gamma_2 + o(t) \end{pmatrix}, \tag{5.6}$$

because, by Lemmas 2.2.5 and 2.2.4 and Proposition 2.2.3,

$$\sum_{i=2}^{n} \frac{b^i_{12}}{a_i^r} = h\delta_{r,\gamma}, \qquad h \neq 0,\ r = 0, \dots, \gamma,\ 2 \leq \gamma \leq n-2. \tag{5.7}$$

Indeed, since the monodromy matrices are nondiagonalizable and since $m \geq \gamma$, it follows that $\omega_{12} = (z-a_l)^{\gamma-1}(1+o(1))$ in (2.30), and $\gamma \leq n-2$, because the representation χ is irreducible (if χ were reducible, then the nondiagonalizability of the monodromy matrices would imply that it is commutative, and it would follow from Proposition 2.2.1 that $\gamma = 0$, which is false). The formula (5.7) now follows from (2.32), (2.31), and (2.30).

The decomposition (1.2.6) for the solution space X of the system (5.3) "at the point" t has the form

$$T_1(y, t) = U_1(z, t)(z-t)^{A_1}(y-t)^{E_1}.$$

Replacing a_i by t and i by 1 in (1.2.9), we get from (1.2.3), (1.2.11), and the equality of the matrices $B^1(0)$ and L_1 that $U_1(t, t)B^1(0) = B^1(t)U_1(t, t)$, and so

$$U_1(t, t) = \begin{pmatrix} 1 + x_1 t + o(t) & o(t) \\ o(1) & 1 + x_2 t + o(t) \end{pmatrix}.$$

Let us reduce the system (5.3) to the form (2.11)–(2.15). To do this we pass from the solution space X of this system to $X' = U^{-1}(t, t)X$. The coefficient matrices of the new system will have the form

$$\widetilde{B}^1(t) = \begin{pmatrix} b + \gamma_1 & 0 \\ 0 & b - \gamma_2 \end{pmatrix},$$

$$\widetilde{B}^i(t) = U_1^{-1}(t, t)B^i(t)U_1(t, t) = \begin{pmatrix} * & b_{12} - (x_1 - x_2)b^i_{12}t - \gamma \frac{b^i_{12}}{a_i} t + o(t) \\ * & * \end{pmatrix}. \tag{5.8}$$

We get from (5.8) and (5.7) that $\sum_{i=2}^n \widetilde{b}^i_{12}(t)/a_i^r = o(t)$, $0 \le r \le \gamma - 2$, and $\sum_{i=2}^n \widetilde{b}^i_{12}(t)/a_i^{\gamma-1} = -\gamma h t + o(t)$, and thus by Lemma 2.2.5 we have for the form $\widetilde{\omega}$ of the new system in a neighborhood of the point $z = t$ that $c^{12}_{\gamma-2}(t) \neq 0$ in (2.31) for sufficiently small t. From this, (2.30), and the equality $\gamma_\omega = \gamma$ we get that $c \neq 0$ and $m \le \gamma - 1$ in the decomposition (2.15) for the new system. We conclude from Proposition 2.2.2 that the Fuchsian weight $\gamma_{\widetilde{\omega}}$ of the new system is less than γ, and therefore $\gamma_{\chi(a')} < \gamma_{\chi(a)}$. It follows from Proposition 2.2.1 that the numbers $\gamma_{\chi(a)}$ and $\gamma_{\chi(a')}$ have the same parity, and thus $\gamma_{\chi(a')} < \gamma_{\chi(a)} - 1$ for a tuple $a' = (a_1 + t, a_2, \dots, a_n)$ for sufficiently small t. To conclude the proof of the proposition it remains only to use the conformal transformation that is the inverse of the one used at the beginning of the proof, and to use Proposition 2.2.5.

It follows from Proposition 2.5.2 and Theorem 2.3.1 that for $p = 3$ all the examples of a negative solution of the RH problem have an unstable character in the sense that the RH problem becomes solvable in the positive sense under an arbitrarily small shift of one of the singular points that preserves the monodromy matrices. Indeed, if the representation (1.1.5) satisfies all the conditions of Theorem 2.3.1, then the Fuchsian weight of the corresponding two-dimensional subrepresentation or factor representation is even (this was proved in Theorem 2.3.2). Therefore, Proposition 2.5.2 gives us that under an arbitrarily small shift of one of the points the Fuchsian weight falls to zero, and hence by Theorem 2.3.1 the corresponding RH problem becomes solvable in the positive sense. What is more, we have the following assertion.

PROPOSITION 2.5.3. *For any tuple* $a = (a_1, \dots, a_n)$ *of points, any representation* (5.2) *of dimension* $p = 3$, *and any* $\varepsilon > 0$ *there is a tuple* $a' = (a'_1, \dots, a'_n)$ *of points such that* $|a_i' - a_i| < \varepsilon$ *and the RH problem has a positive solution for the representation* $\chi(a')$.

Proposition 2.5.3 means, in particular, that there is no collection of 3×3 matrices $G_1, \dots, G_n$ with $G_1 \cdot \dots \cdot G_n = I$ such that the RH problem of the representation $\chi(a)$ has a negative solution for all tuples $a = (a_1, \dots, a_n)$.

6. The Fuchsian weight and two-dimensional holomorphic vector bundles on the Riemann sphere

Recall that in §4 of Chapter I we constructed from any representation (1.1.5) a principal bundle $\widetilde{P}$ and a canonical extension P^0 of this bundle to the whole Riemann sphere. For the vector bundle F^0 associated with the principal bundle P^0 we introduced the concept of the weight $\gamma(0) = \sum_{i=1}^p (c_1^0 - c_i^0)$, where $-c_1^0, \dots, -c_p^0$ are the partial indices of F^0 (see Definition 1.4.2 and the decomposition (1.4.14)).

In §2 of Chapter II we introduced the concept of the Fuchsian weight γ_χ of an arbitrary two-dimensional representation χ (1.1.5) (see Definition 2.2.1). As shown in Theorem 2.3.1, the Fuchsian weight γ_χ is an obstacle to the positive solvability of the RH problem in dimension $p = 3$. The following theorem establishes a connection between the holomorphic bundle F^0 and Fuchsian systems.

THEOREM 2.6.1. *For a representation* χ (1.1.5) *of dimension* $p = 2$ *the weight* $\gamma(0)$ *of the canonical extension* F^0 *of the bundle constructed from the representation* χ *coincides with the Fuchsian weight* γ_χ *of this representation. The partial indices* $-c_1^0$ *and* $-c_2^0$ *of a bundle* F^0 *can be expressed in terms of* γ_χ *by*

$$-c_1^0 = \frac{1}{2}(c_1 - \gamma_\chi) = \frac{1}{2}\left(\sum_{i=1}^n \operatorname{tr} E_i - \gamma_\chi\right),$$
$$-c_2^0 = \frac{1}{2}(c_1 + \gamma_\chi) = \frac{1}{2}\left(\sum_{i=1}^n \operatorname{tr} E_i + \gamma_\chi\right), \tag{6.1}$$

where c_1 *is the first Chern class of the bundle* F^0, *and* E_i *is from* (1.1.9) (*the group* $H^2(CP^1; \mathbb{Z})$ *is identified with* $\mathbb{Z}$).

PROOF. Let us consider the the solution space X of a Fuchsian system (1.1.1) with the given monodromy (1.1.5) and with Fuchsian weight γ_χ. According to Corollary 2.2.1, it can be assumed that this system has the form (2.11)–(2.15), and by Proposition 2.2.2 the number m in (2.15) satisfies the inequality

$$m \geq \gamma_\chi \tag{6.2}$$

(or else the equality $c = 0$ holds).

We construct the bundle P^0 from the representation χ and consider a section $U(z)$ of this bundle:

$$U(z) = W(y)T^{-1}(y), \tag{6.3}$$

where $W(y)$ is from (1.4.3), and $T(y)$ is a fundamental matrix of X. From (2.11) it follows that the section $U(z)$ is holomorphic away from a_l. Since P^0 is holomorphically trivial in the complement of a_l, we have

$$U(z) = (z, U_\infty(z))$$

in the coordinate neighborhood $O_\infty = CP^1 \setminus a_l$, where the matrix-valued function $U_\infty(z)$ is holomorphically invertible in O_∞.

Proceding just as in the proof of necessity in Theorem 1.4.1 (with i replaced by l, φ by 0, and the matrix A^{φ_i} by 0), we get that

$$\begin{aligned}(6.4)\quad U(z) &= (z\,;\, S(y_* - a_l)^{E_l^0} S^{-1}\chi^{-1}(\sigma))\chi(\sigma)S(y_* - a_l)^{-E_l^0}(z-a_l)^{-A_l}U_l^{-1}(z)\\ &= (z\,;\, S(z-a_l)^{-A_l}U_l^{-1}(z)) = (z\,;\, V_l(z))\end{aligned}$$

in the neighborhood O_l of a_l.

The bundle P^0 (and thus the bundle F^0 associated with it) admits the following coordinate description: $O_\infty = CP^1 \setminus a_l$, O_l, $\gamma_{\infty l}$, where $\gamma_{\infty l} = U_\infty(z)V_l^{-1}(z) = U_\infty(z)U_l(z)(z-a_l)^{A_l}S^{-1}$.

But from the form (2.14), (2.15) of the matrices A_l and U_l we get

$$\begin{aligned}U_l(z)(z-a_l)^{A_l} &= \begin{pmatrix} 1 & c(z-a_l)^m \\ s(z-a_l)^k & 1\end{pmatrix}(z-a_l)^{\begin{pmatrix} b+\gamma_1 & 0 \\ 0 & b-\gamma_2\end{pmatrix}}(1+o(1))\\ &= (z-a_l)^{\begin{pmatrix} b+\gamma_1 & 0 \\ 0 & b-\gamma_2\end{pmatrix}} \cdot \begin{pmatrix} 1 & c(z-a_l)^{m-\gamma_\chi} \\ s(z-a_l)^{k+\gamma_\chi} & 1\end{pmatrix}(1+o(1)).\end{aligned}$$

In view of the condition (6.2) (and, of course, in the case $c = 0$) we get that the matrix

$$\widetilde{U}_l(z) = \begin{pmatrix} 1 & c(z-a_l)^{m-\gamma_\chi} \\ s(z-a_l)^{k+\gamma_\chi} & 1\end{pmatrix}(1+o(1))$$

is holomorphically invertible in some neighborhood O_l' of a_l; therefore,

$$\gamma_{\infty l} = U_\infty(z)(z-a_l)^{A_l}\widetilde{U}_l(z)S^{-1}.$$

This last means that the bundle F^0 is holomorphically equivalent to the bundle with neighborhoods O_∞ and O_l' and transition function

$$\widetilde{\gamma}_{\infty l} = (z-a_l)^{\begin{pmatrix} b+\gamma_1 & 0 \\ 0 & b-\gamma_2\end{pmatrix}},$$

but the latter bundle is holomorphically equivalent to the bundle

$$O(-(b+\gamma_1)) \oplus O(-(b-\gamma_2)),$$

and hence

$$-c_1^0 = -b-\gamma_1, \qquad -c_2^0 = -b+\gamma_2, \qquad \gamma(0) = c_1^0 - c_2^0 = \gamma_\chi.$$

The equalities (6.1) now follow from Remark 1.4.2.

COROLLARY 2.6.1. *Any two-dimensional holomorphic vector bundle on* CP^1 *is holomorphically equivalent to the bundle* F^0 *that is the canonical extension of the bundle constructed from the irreducible representation* (1.1.5).

PROOF. This follows from (1.4.14), Proposition 2.2.4, and Theorem 2.6.1.

From Theorem 2.6.1 we get the following formulation of Theorem 2.3.1.

THEOREM 2.6.2. *The Riemann–Hilbert problem for a representation* (1.1.5) *of dimension* $p = 3$ *has a negative solution if and only if*:

1) *all the local representations* (1.1.7) *constructed from* (1.1.5) *are indecomposable*;

2) *the representation* (1.1.5) *is reducible, and the corresponding two-dimensional subrepresentation or factor representation* χ^2 *is irreducible*;

3) *the partial indices of the canonical extension of the bundle constructed from the representation* χ^2 *are distinct.*

REMARK 2.6.1. As shown in Remark 2.3.2, the condition of irreducibility of χ^2 follows from the remaining conditions. However, we included it in the statement of the theorem because, in contrast to the condition 3), it is easy to verify. We remark also that Theorem 2.6.2 is a reformulation of Theorem 1.4.2 for the case $p = 3$. It turns out that in this case all the examples of a negative solution of the RH problem are given by this theorem.

In any complex vector bundle constructed from a monodromy representation it is possible to introduce a holomorphic integrable connection (see [31]). Let $(e_1^i, \dots, e_p^i)$ be an arbitrary basis of sections of the bundle $\widetilde{F}|_{O_i}$, where $\widetilde{F}$ denotes the vector bundle constructed in §4 of Chapter I. Denote by $\widetilde{\nabla}$ a holomorphic connection in $\widetilde{F}$, and by $-\omega^i$ the matrix of this connection in the basis $(e_1^i, \dots, e_p^i)$. Let f be the coordinates of the section s in the basis (e^i). Then

$$\widetilde{\nabla} f = (d - \omega^i) f.$$

Thus, the connection $\widetilde{\nabla}$ gives a family of local systems of differential equations on $CP^1 \setminus \mathscr{D}$:

$$(d - \omega^i) f = 0. \tag{6.5}$$

In a basis $(s_1, \dots, s_p)$ of p linearly independent (at each point) global holomorphic sections, which exists in view of the holomorphic triviality of the bundle $\widetilde{F}$, the family (6.5) gives a system (1.1.1) of linear differential equations on the whole of $CP^1 \setminus \mathscr{D}$. If we pass from the basis $(s_1, \dots, s_p)$ to $(s'_1, \dots, s'_p) = (s_1, \dots, s_p)\Gamma^{-1}(z)$, then the system (1.1.1) constructed from $\widetilde{\nabla}$ becomes

$$df = \omega' f,$$

where ω' has the form (1.3.2). Thus, specifying a connection in the trivial bundle is equivalent to specifying a system of linear differential equations (to within the gauge transformations (1.3.1)).

We consider all possible extensions of the bundle $(\widetilde{F}, \widetilde{\nabla})$ to the whole of CP^1 for which the connection has singularities of simple pole type at the points of $\mathscr{D}$. It follows from Theorem 1.4.1 that all such extensions $(F^\varphi, \widetilde{\nabla})$ coincide with the extensions constructed from all possible admissible tuples φ.

It follows from Corollary 2.6.1 that in any two-dimensional holomorphic vector bundle E on CP^1 it is possible to introduce an irreducible connection ∇^φ that is holomorphic away from a finite set of points and has singularities of simple pole type at these points; we call such connections *irreducible Fuchsian connections.*

Denote by n^φ the number of singular points of the irreducible Fuchsian connection ∇^φ. Consider the set Φ the set of all such connections, and let $n = \min_\Phi n^\varphi$.

COROLLARY 2.6.2. *The number* n *is connected with the partial indices* $-c_1^E$ *and* $-c_2^E$ *of a bundle* E *by the following relations (for* $c_1^E \neq c_2^E$):

$$n - 2 = c_1^E - c_2^E.$$

The proof follows from Propositions 2.2.3 and 2.2.4 and Theorem 2.6.1. Thus, the holomorphy type (c_1^E, c_2^E) of a bundle on CP^1 with $c_1^E \neq c_2^E$ is completely determined by its Chern class $-(c_1^E + c_2^E)$ (a topological invariant) and by the minimal number $n = c_1^E - c_2^E + 2$ of singular points of all possible irreducible Fuchsian connections for the given bundle (an analytic invariant).

We consider a finite covering $\mathfrak{A}$ of the Riemann sphere CP^1. Let $\mathfrak{M}$ be a covering with $\mathfrak{M} \ll \mathfrak{A}$. The holomorphic bundle E on CP^1 is determined by a cocycle

$$f_E \in \mathbb{Z}^1(\mathfrak{A}; O^{GL(p;C)}).$$

From Theorem 2.6.1 and Proposition 2.5.2 we get a known result about the instability of the partial indices (see [39]).

PROPOSITION 2.6.1. *For any two-dimensional holomorphic vector bundle* E *on* CP^1 *with first Chern class* c_1, *and for any* $\varepsilon > 0$, *there is a holomorphic vector bundle* E' *such that* $\|f_E - f_{E'}\|_{C(\mathfrak{M})} < \varepsilon$, *and the partial indices* $-c_1^{E'}$ *and* $-c_2^{E'}$ *of the bundle* E' *are equal to*

$$-c_2^{E'} = \left[\frac{c_1 + 1}{2}\right], \qquad -c_1^{E'} = c_1 + c_2^{E'},$$

where $[x]$ *is the integer part of a number* x.

CHAPTER III

Fuchsian Equations and Fuchsian Systems on the Riemann Sphere

1. The number of false singular points of a Fuchsian equation

We recall that a pth-order Fuchsian equation at a point a_l is defined to be a linear differential equation

$$u^{(p)} + q_1(z)u^{(p-1)} + \cdots + q_p(z)u = 0 \tag{1.1}$$

whose coefficients in a neighborhood O_l of a_l have the form

$$q_i(z) = \frac{r_i(z)}{(z-a_l)^i}, \qquad i = 1, \ldots, p, \tag{1.2}$$

where the $r_i(z)$ are holomorphic functions on O_l.

Equation (1.1) is said to be Fuchsian on the whole Riemann sphere if it has there finitely many singular points and is Fuchsian at each such point.

Let $\mathscr{D} = \{a_1, \ldots, a_n\}$ be the set of singular points of equation (1.1) on the Riemann sphere. It can be assumed without loss of generality that $a_1 = 0$ and that the point ∞ is not among the singular points (these conditions can always be achieved by a conformal transformation of the Riemann sphere). The latter means that with respect to the local coordinate $\zeta = 1/z$ in a neighborhood of ∞ $(\zeta = 0)$ the coefficients $\widetilde{q}_i(\zeta)$ of the equation

$$\frac{d^p u}{d\zeta^p} + \widetilde{q}_1(\zeta)\frac{d^{p-1}u}{d\zeta^{p-1}} + \cdots + \widetilde{q}_p(\zeta)u = 0 \tag{1.3}$$

obtained from (1.1) by the substitution $z = 1/\zeta$ are holomorphic at $\zeta = 0$.

Since

$$\frac{d^{j-1}u}{dz^{j-1}} = \frac{1}{z^{2(j-1)}}\sum_{i=2}^{j} c_i^j z^{j-i}\frac{d^{i-1}u}{d\zeta^{i-1}}, \tag{1.4}$$

where $c_j^i = (-1)^{j-1}$, the $\widetilde{q}_i(\zeta)$ are holomorphic at $\zeta = 0$ for $i = 1, \ldots, p$ if and only if the functions

$$\widetilde{r}_i(\zeta) = \frac{\zeta^{2i}}{\zeta^{2p}} q_i\left(\frac{1}{\zeta}\right), \qquad i = 1, \ldots, p, \tag{1.5}$$

are holomorphic at $\zeta = 0$.

We get from (1.2) and (1.5) that each $q_i(z)$ is a rational function, and

$$q_i(z) = \frac{P_i(z)}{\prod_{l=1}^{n}(z-a_l)^i}, \tag{1.6}$$

where $P_i(z)$ is a polynomial of degree k_i, and k_i satisfies the equality

$$(1.7)\qquad 2i + ni \geq 2p + k_i, \qquad k_i \leq (n+2)i - 2p.$$

The k_ith-degree polynomial P_i in (1.6) contains $k_i + 1$ independent parameters (the coefficients of the polynomial); therefore, an arbitrary Fuchsian equation on CP^1 with singularities on $\mathscr{D}$ depends on

$$(1.8)\qquad N_{\text{eq}} = \sum_{i=1}^{p}(k_i+1) = \frac{(n+2)p(p+1)}{2} - 2p^2 + p = \frac{p^2}{2}(n-2) + \frac{pn}{2}$$

independent parameters.

An arbitrary irreducible representation (1.1.5) depends on

$$(1.9)\qquad N_{\text{rep}} = p^2(n-2) + 1$$

independent parameters. Indeed, in view of the relation $G_1 \cdot \ldots \cdot G_n = I$ the representation (1.1.5) is given by the $n-1$ matrices $G_1, \ldots, G_{n-1}$, which involve $p^2(n-1)$ parameters. Since the representation is determined up to equivalence, each G_i is specified up to $S^{-1}G_iS$. If $S_1^{-1}G_iS_1 = S_2^{-1}G_iS_2$ for all $i = 1, \ldots, n-1$, then according to Schur's lemma $S_1S_2^{-1} = \lambda I$ for some $\lambda \in C$, and hence the number of parameters should be decreased by $p^2 - 1$. Finally, we get

$$N_{\text{rep}} = p^2(n-1) - (p^2-1) = p^2(n-2) + 1.$$

The difference Δ between the numbers N_{rep} and N_{eq} is

$$(1.10)\qquad \Delta = N_{\text{rep}} - N_{\text{eq}} = \frac{p(p-1)}{2}(n-2) - p + 1.$$

It follows from (1.10) that $\Delta > 0$ for $n > 3$ and any $p > 1$. For $n = 3$ the number Δ is equal to 0 only for $p = 2$.

The inequality $\Delta > 0$ noted above means that for $n > 3$ and $p > 1$ (or for $n = 3$ and $p > 2$) the construction of a Fuchsian equation (1.1) with a given monodromy (1.1.5) is possible in the general case only when there are additional, so-called false singular points $b_1, \ldots, b_m$ that do not contribute to the monodromy.

In this section we obtain an upper estimate for the number of such points as a function of the weight $\gamma(0)$ of the bundle F^0 constructed from an irreducible representation (1.1.5) (see Definition 1.4.2).

THEOREM 3.1.1. *For any irreducible representation* (1.1.5) *there exists a Fuchsian equation* (1.1) *with given monodromy* (1.1.5) *such that the number* m *of additional false singular points satisfies the inequality*

$$(1.11)\qquad m \leq \frac{p(p-1)}{2}(n-2) - \gamma(0) + 1 - l,$$

where l *is the number of partial indices in the decomposition* (1.4.14) *for the bundle* F^0 *that are equal to the first partial index.*

PROOF. According to Proposition 1.4.1, there exists a system (1.1.1) with regular singular points $a_1, \ldots, a_n$ and the given monodromy (1.1.5) that has Fuchsian singularities at $a_1, \ldots, a_n$ and is such that all the normalizations at $a_2, \ldots, a_n$ are equal to zero. Let $T(y)$ be a fundamental matrix of the solution space X of (1.1.1).

Denote by $u_1(y), \dots, u_p(y)$ the elements of some row of $T(y)$. If the representation (1.1.5) is irreducible, then the analytic functions $u_1(y), \dots, u_p(y)$ are linearly independent, and hence the Wronskian $W(y)$ of these functions is not identically equal to zero. Therefore, $u_1(y), \dots, u_p(y)$ form a basis in the solution space Y of the Fuchsian equation

$$\frac{1}{W(y)}\begin{vmatrix} u & u_1(y) & \cdots & u_p(y) \\ \frac{du}{dz} & \frac{du_1(y)}{dz} & \cdots & \frac{du_p(y)}{dz} \\ \cdots & \cdots & \cdots & \cdots \\ \frac{d^pu}{dz^p} & \frac{d^pu_1(y)}{dz^p} & \cdots & \frac{d^pu_p(y)}{dz^p} \end{vmatrix} = 0. \tag{1.12}$$

It follows from the construction that the monodromy of equation (1.12) coincides with the given monodromy, and the additional false singular points of this equation are the projections of the zeros of $W(y)$. The proof of the theorem reduces to an estimate of the number of projections of the zeros of this Wronskian.

We consider an equation (1.1) that is Fuchsian at the point a_l.

LEMMA 3.1.1. *Suppose that the function $h(z)$ is holomorphic in the neighborhood O_l of a_l, and $h(a_l) \neq 0$. The system* (1.1.1) *constructed from equation* (1.1) *with the help of the substitution*

$$f^j(y) = ((z - a_l)h(z))^{j-1}\frac{d^{j-1}u}{dz^{j-1}}, \tag{1.13}$$

is Fuchsian at a_l, and its exponents β_l^j at this point coincide with those of equation (1.1) (*that is, with the roots of the corresponding defining equation*).

PROOF. The matrix form ω of the system (1.1.1) constructed with the help of the substitution (1.13) from equation (1.1) is

(1.14)
$$\omega = \frac{dz}{z-a_l}\begin{pmatrix} 0 & \frac{1}{h} & 0 & \cdots & & 0 \\ & & & & & 0 \\ 0 & \frac{((z-a_l)h)'}{h} & \frac{1}{h} & & \ddots & \vdots \\ \vdots & & \ddots & \ddots & & \\ 0 & & & & \frac{1}{h} & 0 \\ 0 & 0 & \cdots & & \frac{(p-1)((z-a_l)h)'}{h} & \frac{1}{h} \\ -r_ph^{p-1} & -r_{p-1}h^{p-2} & \cdots & & -r_2h & \frac{-r_1h-(p-1)((z-a_l)h)'}{h} \end{pmatrix};$$

therefore, the system (1.1.1) is Fuchsian at a_l. Let $u_1, \dots, u_p$ be a basis for the solution space Y of (1.1) that is associated in $Y|_{\widetilde{O}_l}$. (An associated basis in the solution space of an equation is defined and constructed in precisely the same way as it was in the solution space of a system in §1 of Chapter I.) Then the basis constructed from it with the help of (1.13) in the solution space X of the system (1.1.1) is also associated, because it follows from (1.13) that

$$\varphi_l(f_i) = \varphi_l(u_i), \qquad i = 1, \dots, p,$$

for the normalizations φ_l at a_l. Therefore, the exponents β_l^j of the solution space of the constructed system at a_l coincide with the exponents of equation (1.1) at this point. We recall here that the exponents β_l^j of (1.1) at a_l are the roots of the defining equation

(1.15) $\lambda(\lambda-1)\cdot\dots\cdot(\lambda-p+1)+r_1(a_l)\lambda(\lambda-1)\cdot\dots\cdot(\lambda-p+2)+\dots+r_p(a_l)=0$.

If β_l^j is a root of equation (1.15) with multiplicity k, then equation (1.1) has k linearly independent solutions of the form

$$ {}^ju_i(y)=(y-a_l)^{\beta_i^j}\left(\sum_{t=1}^{i}\varphi_t(z)\ln^{i-t}(y-a_l)\right), \tag{1.16}$$

$$i=1,\dots,k, \qquad \varphi_1(a_l)\neq 0,$$

where $\varphi_1,\dots,\varphi_k$ are holomorphic in O_l (see [1], [12]). It follows immediately from the form of the solutions (1.16) that the numbers β_l^j are the exponents of the solution space Y of (1.1) in the sense of Definition 1.2.1.

LEMMA 3.1.2. *The system* (1.1.1) *constructed from a Fuchsian equation with singular points* $a_1,\dots,a_n$ *by means of the substitution*

$$ f^j(y)=\left(\prod_{i=1}^{n}(z-a_i)\right)^{j-1}\frac{d^{j-1}u}{dz^{j-1}}, \tag{1.17}$$

is Fuchsian at $a_1,\dots,a_n$ *and has the single additional false singular point* ∞, *at which all the regular branches of the determinant* $\det T(y)$ *of the fundamental matrix* $T(y)$ *of the solution space of this system have a pole of order* $(n-2)p(p-1)/2$.

PROOF. The Fuchsian property of the constructed system at $a_1,\dots,a_n$ follows from Lemma 1.1.2. Since ∞ is not a singular point for equation (1.1), each regular branch $R(\zeta)$ of the matrix

$$\begin{pmatrix} u_1 & \cdots & u_p \\ \frac{du_1}{d\zeta} & \cdots & \frac{du_p}{d\zeta} \\ \cdots & \cdots & \cdots \\ \frac{d^{p-1}u_1}{d\zeta^{p-1}} & \cdots & \frac{d^{p-1}u_p}{d\zeta^{p-1}} \end{pmatrix}, \tag{1.18}$$

where $u_1,\dots,u_p$ is a basis for the solution space of (1.1), is holomorphically invertible at the point $\zeta=0$ ($\zeta=1/z$ is the coordinate in a neighborhood of ∞). By (1.4), the fundamental matrix $T(y)$ of (1.1.1) constructed according to (1.1.7) has in a neighborhood of any point in $\pi^{-1}(\infty)$ the form

$$ T(y)=\Gamma_2(z)\Gamma_1(z)R(1/z), \tag{1.19}$$

where $R(1/z)$ is some branch of the matrix (1.18), and where, in turn, $\Gamma_1(z)=\|\gamma_{ij}(z)\|$ is a lower triangular matrix with elements

$$ \gamma_{11}=1, \qquad \gamma_{j1}=0, \quad j\neq 1, \qquad \gamma_{ji}=c_i^jx^{j-i}, \quad 1<i\leq j, \tag{1.20}$$

$$ \Gamma_2(z)=\operatorname{diag}\left(1,\frac{1}{z^2}\prod_{l=1}^{n}(z-a_l),\dots,\frac{1}{z^{2(p-1)}}\prod_{l=1}^{n}(z-a_l)^{p-1}\right). \tag{1.21}$$

The last part of the lemma now follows from the holomorphic invertibility of the matrix $R(1/z)$ and the fact that

$$\det\Gamma_1(z)\equiv\prod_{j=2}^{n}c_j^i\neq 0 \quad\text{and}\quad \det\Gamma_2(z)=z^{(n-2)p(p-1)/2}h(z),$$

where the function $h(z)$ is holomorphically invertible at ∞. We return to the system (1.1.1) constructed at the beginning of the proof of the theorem from the irreducible representation (1.1.5).

Consider the decompositon (1.2.6) for the solution space X of this system on $\widetilde{O}_1$. Since the normalizations φ_i^j of the solution space of (1.1.1) are equal to zero for all $i = 2, \ldots, n$ and $j = 1, \ldots, p$, we get exactly as in the proof of Theorem 2.6.1 that the bundle F^0 admits the following coordinate description: $O_\infty = CP^1 \setminus a_1$, O_1, $\gamma_{\infty 1} = U_1(z)(z-a_1)^{A_1}$. It follows from Lemma 1.3.9 that there exist a matrix $\Gamma(z)$ holomorphically invertible away from a_1 and a function $V_1(z)$ holomorphically invertible in the neighborhood O_1 of a_1 such that

$$\Gamma(z)U_1(z)(z-a_1)^{A_1} = (z-a_1)^C V_1(z), \tag{1.22}$$

where $C = \operatorname{diag}(k_1, \ldots, k_p)$, $k_1, \ldots, k_p \in \mathbb{Z}$. According to Lemma 1.3.10, the numbers $k_1, \ldots, k_p$ coincide with the numbers $c_1^0, \ldots, c_p^0$ in the decomposition (1.4.14) for F^0.

We pass from the system (1.1.1) with solution space X to a system with the solution space $X' = \Gamma(z)X$. Here the decompositions (1.2.6) for the space X' will have the same form as for X, with the matrices $U_i(z)$ replaced by $\Gamma(z)U_i(z)$.

Suppose that $k_1 = \cdots = k_l$ in (1.22). There is a nonsingular constant matrix

$$S = \begin{pmatrix} \widetilde{S} & 0 \\ 0 & I_{p-1} \end{pmatrix} \tag{1.23}$$

such that the following equalities hold for the first row $(v_1'(z), \ldots, v_p'(z))$ of the matrix $V_1' = SV_1$:

$$v_1'(a_1) = \cdots = v_{l-1}'(a_1) = 0. \tag{1.24}$$

We pass from the system (1.1.1) with solution space X' to a system with solution space $X'' = SX'$, denoting the latter system again by (1.1.1), and the space X'' by X.

Consider the elements $u_1(y), \ldots, u_p(y)$ of the first row of some fundamental matrix of the solution space of the system constructed. The space Y of functions generated by $u_1, \ldots, u_p$ is the solution space of the Fuchsian equation (1.12) with given monodromy (1.1.5). It can be assumed without loss of generality that ∞ is not among the singular points of (1.12), and therefore the set $\{b_1, \ldots, b_m\}$ of additional false singular points of this equation coincides with the set of projections of the zeros of the Wronskian $W(y)$ in the finite plane, and the latter set coincides with the set of projections of the zeros of the determinant of the fundamental matrix $T(y)$ constructed in Lemma 3.1.2 from the solutions $u_1, \ldots, u_p$ with the help of (1.17). We consider the form $d \ln \det T(y)$ on CP^1. Denote by p_i the residues of this form at the points $a_1, \ldots, a_n$, and by d_j the residues at the points $b_1, \ldots, b_m$; then $d_j > 0$ for all j.

Lemma 3.1.1, the decomposition (1.2.6), and the fact that $\varphi(f) \le \varphi(f^k)$ ($k = 1, \ldots, p$) for the normalization $\varphi(f)$ of any vector-valued function $f = (f^1, \ldots, f^p)$

give us

$$p_i = \sum_{j=1}^{p} \beta_i^j + s_i, \qquad s_i \geq 0, \quad i = 2, \ldots, n \tag{1.25}$$

(the number s_i can be greater than zero; for example, if the element ${}^i u_1^1(z)$ of the matrix $U_i(z)$ has a zero of order r at a_i, then $s_i \geq r$). By (1.22) and (1.24), we have at a_1 that

$$p_1 = \sum_{j=1}^{p} \rho_1^j + pk_1 + (l-1) + s_1, \qquad s_1 \geq 0. \tag{1.26}$$

We get from the residue theorem and Lemma 3.1.2 that

$$\sum_{j=1}^{p} \rho_1^j + \sum_{i=2}^{n}\sum_{j=1}^{p} \beta_i^j + pk_1 + \sum_{i=1}^{n} s_i + (l-1) + \sum_{k=1}^{m} d_k = \frac{(n-2)p(p-1)}{2}. \tag{1.27}$$

On the other hand, (1.2.6) and the decomposition (1.22) give us

$$\sum_{i=2}^{n}\sum_{j=1}^{p} \beta_i^j + \sum_{j=1}^{p} \rho_1^j + \sum_{i=1}^{p} k_i = 0$$

by the residue theorem for the form $d \ln \det T_i(y)$, where $T_i(y)$ is a fundamental matrix of the solution space of the original system (1.1.1); therefore, (1.27) takes the form

$$\sum_{k=1}^{m} d_k = \frac{(n-2)p(p-1)}{2} - \left(\sum_{i=1}^{p}(k_1 - k_i) + l - 1 + \sum_{i=1}^{n} s_i\right). \tag{1.28}$$

Since the number m of additional false singular points does not exceed $\sum_{k=1}^{m} d_k$, we get from (1.28) that

$$m \leq \frac{(n-2)p(p-1)}{2} - \left(\sum_{i=1}^{p}(k_1 - k_i) + l - 1 + \sum_{i=1}^{n} s_i\right). \tag{1.29}$$

The theorem now follows from the fact that $s_i \geq 0$, $i = 1, \ldots, n$.

COROLLARY 3.1.1. *For $n = 3$ any irreducible representation* (1.1.5) *of dimension $p = 2$ can be realized as the monodromy of the Gaussian equation* (*that is, of a second-order Fuchsian equation with three singular points*).

COROLLARY 3.1.2. *In the case of an irreducible representation* (1.1.5) *the numbers $k_1, \ldots, k_p$ for the bundle F^0 satisfy*

$$\gamma(0) = \sum_{i=1}^{p}(k_1 - k_i) \leq \frac{(n-2)p(p-1)}{2} + 1 - l. \tag{1.30}$$

If the inequality (1.30) is an equality, then the representation (1.1.5) can be realized as the monodromy representation of a Fuchsian equation without additional false singular points.

REMARK 3.1.1. If $u_1, \ldots, u_p$ is a basis in the solution space of a Fuchsian equation (1.1) with singular points $a_1, \ldots, a_n$, then, repeating the arguments of Lemma 3.1.2 for this basis, we get instead of (1.27) the classical Fuchs relation

$$\sum_{i=1}^{n}\sum_{j=1}^{p}\beta_i^j = \frac{(n-2)p(p-1)}{2}, \tag{1.31}$$

where the β_i^j are the exponents of equation (1.1) at the point a_i, $j = 1, \ldots, p$.

REMARK 3.1.2. If in the proof of the theorem we consider instead of (1.22) the decomposition

$$\Gamma'(z)U_1(z) = (z-a_1)^{C'}V_1'(z) \tag{1.22$'$}$$

(the existence of which was proved in Lemma 1.3.9), where $C' = \operatorname{diag}(c_1, \ldots, c_p)$, $c_i \in \mathbb{Z}$, $c_1 \geq \cdots \geq c_p \geq 0$, then the formulas (1.29) and (1.11) will also hold with the numbers k_i replaced by c_i. The proof almost completely repeats that of the theorem with the formula (1.26) replaced by

$$p_1 = \sum_{j=1}^{p}\beta_1^j + pc_1 + l - 1 + s_1 \tag{1.26$'$}$$

and the relations

$$\sum_{i=2}^{n}\sum_{j=1}^{p}\beta_i^j + \sum_{j=1}^{p}\rho_1^j + \sum_{i=1}^{p}k_i = 0$$

replaced by

$$\sum_{i=1}^{n}\sum_{j=1}^{p}\beta_i^j + \sum_{i=1}^{p}c_i = 0.$$

REMARK 3.1.3. Theorem 3.1.1 improves the estimate

$$m \leq \frac{(n-2)p(p-1)}{2} + 1 - p \tag{1.32}$$

obtained in [51] for the case when one of the local monodromy representations is diagonalizable (the two estimates coincide for $\gamma(0) \leq 1$).

2. Construction from a Fuchsian equation of a Fuchsian system with the same singular points and the same monodromy

It is known that from any equation (3.1.1) that is Fuchsian in the neighborhood O_i of the point a_i it is possible to construct a Fuchsian system in O_i with the help of the substitution $f^j = ((z-a_i)\frac{d}{dz})^{j-1}u$ [16], but this construction bears an essentially local character. In this section we prove the following assertion.

THEOREM 3.2.1. *From any Fuchsian equation* (1.1) *on the Riemann sphere it is possible to construct a Fuchsian system*

$$df = \left(\sum_{i=1}^{n}\frac{B^i}{z-a_i}dz\right)f \tag{2.1}$$

with the same singular points and the same monodromy.

Theorem 3.2.1 answers a question raised by Erugin [14]. Moreover, it enables us to give examples of representations (1.1.5) that trivially are not realizable as monodromy representations of Fuchsian equations (1.1) with additional false singular points.

Proof. Consider an equation (1.1) with solution space Y that is Fuchsian in the neighborhood O_l of a_l, and choose an associated basis $u_1, \dots, u_p$ in $Y|_{\tilde{O}_l}$. The latter means that, in particular, the fundamental matrix $T(y)$ constructed from $u_1, \dots, u_p$ with the help of the substitution (1.17) has a decomposition of the form (1.2.6) that satisfies the condition (1.2.8) for $i = l$ (see Lemma 3.1.1).

Lemma 3.2.1. *All the principal minors of the matrix $U_l(z)$ in (1.2.6) for the fundamental matrix $T(y)$ are nonzero at a_l.*

Proof. Consider the first i elements $u_1, \dots, u_i$ of an associated basis for the space $Y|_{\tilde{O}_l}$. The analytic functions $u_1(y), \dots, u_i(y)$ are linearly independent, and the space Y_i generated by them is invariant under the action of the monodromy operator (this follows from the fact that the matrix E_l in the decomposition (1.2.6) is upper triangular). Therefore, Y_i is the solution space of some equation (1.12) that is Fuchsian in O_l. From the uniqueness theorem for the Wronskian $W_i(y)$ of the functions $u_1, \dots, u_i$, multiplied by $(y-a_l)^{-\operatorname{tr} E_l^i}$, where E_l^i is the matrix formed by the first i columns and rows of E_l (note that the function obtained is meromorphic at a_l), it follows that the number of additional false singular points of this equation does not have a_l as a limit point. We reduce the size of O_l to a neighborhood O_l' that does not contain any of these points. The assertion of the lemma now follows from the fact that the ith principal minor of the matrix $U_l(z)$ in the lemma coincides with the determinant $\det U_l^i(z)$ of the matrix $U_l^i(z)$ in the decomposition (1.2.6) for the fundamental matrix $T^i(y)$ constructed from $u_1, \dots, u_i$ with the help of (1.17), and this determinant is nonzero at a_l according to Lemma 3.1.1 and (1.2.8).

Let us return to the proof of the theorem. It can be assumed without loss of generality that $a_1 = 0$ and that the point ∞ is not among the singular points of (1.1) (these conditions can always be achieved by a conformal transformation of the Riemann sphere). With the help of the substitution (1.17) we pass from the Fuchsian equation (1.1) to a system (1.1.1), and we denote its solution space by X. According to Lemma 3.1.2, this system is Fuchsian at $a_1, \dots, a_n$, and the additional false singular point ∞ will be a regular singular point for it. The decomposition (1.19) holds for the fundamental matrix $T(y)$ of the solution space of this system. Denote by $\Gamma_3(z)$ and $\Gamma_4(z)$ the matrices

$$\Gamma_3(z) = \operatorname{diag}(1, z^{-(n-2)}, \dots, z^{-(n-2)(p-1)}), \qquad \Gamma_4(z) = z^{(n-2)-1}\Gamma_3, \tag{2.2}$$

and by $\Gamma_5(z)$ the matrix

$$\Gamma_5(z) = \Gamma_3\Gamma_2\Gamma_1\Gamma_2^{-1}\Gamma_3^{-1}. \tag{2.3}$$

It follows from the form (1.20) of the matrix $\Gamma_1(z)$ and, in particular, from its lower triangularity that the matrix $\Gamma_5(z)$ is lower triangular and holomorphically invertible away from the points $a_1 = 0$ and ∞, and that the elements of its main diagonal are constant nonzero numbers (equal to the numbers c_j^i in (1.20)). Therefore, the same is true also for the matrix Γ_5^{-1}.

Lemma 3.2.2. *There exists a lower triangular meromorphic matrix $\Gamma_6(z)$ on CP^1 that is holomorphically invertible away from the point $a_1 = 0$ and such that the matrix*

$\Gamma_7(z) = \Gamma_6(z)\Gamma_5^{-1}$ *is holomorphically invertible away from the point* ∞ *and has lower triangular form with ones on the main diagonal.*

PROOF. Let $\Gamma_5^{-1}(z) = ((\gamma_{ji}(z)))$ and $S = \operatorname{diag}(\gamma_{11}, \dots, \gamma_{pp})$. Then ${}^0\Gamma_7(z) = S^{-1}\Gamma_5^{-1}(z)$ has ones on the main diagonal and is meromorphic at zero. Any element ${}^0\gamma_{ji}$ of it has the form ${}^0\gamma_{ji}(z) = {}^0Q^{ji}(1/z) + {}^0h_{ji}(z)$, where ${}^0Q^{ji}(1/z)$ is a polynomial, and the function ${}^0h_{ji}(z)$ is holomorphic at zero. Therefore, the $(p-1)$st and pth columns of the matrix ${}^1\Gamma_7 = {}^{p-1}\Gamma_6\Gamma_7^0$ are holomorphic at zero, where

$$ {}^{p-1}\Gamma_6 = \begin{pmatrix} 1 & & & 0 \\ & \ddots & & \\ & & 1 & \\ 0 & & -{}^0Q^{pp-1} & 1 \end{pmatrix}. $$

The remaining elements ${}^1\gamma_{ji}$ of it again have the form ${}^1\gamma_{ji}(z) = {}^1Q^{ji}(1/z) + {}^1h_{ji}(z)$. Therefore, the $(p-2)$nd, $(p-1)$st, and pth columns of the matrix ${}^2\Gamma_7 = {}^{p-2}\Gamma_6{}^1\Gamma_7$ are holomorphic, where

$$ {}^{p-2}\Gamma_6 = \begin{pmatrix} 1 & & & & 0 \\ & \ddots & & & \\ & & 1 & & \\ 0 & & -{}^1Q^{p-1p-2} & 1 & 0 \\ & & -{}^1Q^{pp-2} & 0 & 1 \end{pmatrix}, $$

and so on.

Use of the procedure whose first two steps were just described leads in $p-2$ steps to the required matrices $\Gamma_6 = (\prod_{i=2}^{p-1} {}^i\Gamma_6)S^{-1}$ and $\Gamma_7(z) = \Gamma_6(z)\Gamma_5^{-1}(z)$. The lemma is proved.

The matrix $\Gamma_7(z)\Gamma_4(z)$ can be written as

$$ \Gamma_7(z)\Gamma_4(z) = \Gamma_4(z)\Gamma_8(z), \tag{2.4} $$

where $\Gamma_8(z) = \Gamma_4^{-1}\Gamma_7\Gamma_4$ is a lower triangular matrix that is holomorphically invertible away from $a_1 = 0$ and meromorphic at ∞ and that has ones on the main diagonal.

We pass from the system (1.1.1) with solution space X constructed with the help of the substitution (1.17) to a system (1.1.1) with the solution space

$$ X' = \Gamma_4(z)\Gamma_8(z)z^{-(n-2)(p-1)}X. \tag{2.5} $$

LEMMA 3.2.3. *The system* (1.1.1) *with solution space* X' *is Fuchsian at the points* $a_2, \dots, a_n$, *the point* $a_1 = 0$ *is a regular singular point for it, and the point* ∞ *is a point of holomorphy.*

PROOF. The Fuchsian property of the system at $a_2, \dots, a_n$ follows from the fact that $\det(\Gamma_4(a_i)\Gamma_8(a_i)a_i^{-(n-2)(p-1)}) \neq 0$ and from (1.2.8). We get from (1.20), (1.21), and (2.2)–(2.4) that

$$ \begin{aligned} \Gamma_4\Gamma_8 z^{-(n-2)(p-1)}T(y) &= \Gamma_7\Gamma_3 T(y) = \Gamma_6\Gamma_5^{-1}\Gamma_3 T(y) \\ &= \Gamma_6\Gamma_3\Gamma_2\Gamma_1^{-1}\Gamma_2^{-1}\Gamma_3^{-1}T(y) = \Gamma_6\Gamma_3\Gamma_2 R(1/z), \end{aligned} $$

but the matrices $\Gamma_6(z)$, $\Gamma_3(z)\Gamma_2(z)$, and $R(1/z)$ are holomorphically invertible at ∞. Therefore, this point is not a singular point for the system constructed. The lemma is proved.

The decomposition (1.2.6) for the space $X'|_{\tilde{O}_1}$ is connected with the analogous decomposition for X as follows:

$$T_1'(y) = \Gamma_4()U_1^0(z)z^{A_1'}y^{E_1} = z^K U_1^0 z^{A_1'} y^{E_1}, \tag{2.6}$$

where $K = \operatorname{diag}((n-2)(p-1), \dots, n-2, 0)$, $U_1^0(z) = \Gamma_8(z)U_1(z)$, and $A_1' = A_1 - (n-2)(p-1)I$. Since $\Gamma_8(z)$ is a lower triangular matrix holomorphically invertible at zero, Lemma 3.2.1 tells us that all the principal minors of $U_1^0(z)$ are nonzero at $a_1 = 0$. It follows from the form of the matrix A_1' and Corollary 1.2.1 that the normalizations ${}'\varphi_1^j$ of X' at zero are connected with the corresponding normalizations φ_1^j of X by the relations

$${}'\varphi_1^j \geq \varphi_1^j - (n-2)(p-1), \qquad j = 1, \dots, p. \tag{2.7}$$

LEMMA 3.2.4. *There is a meromorphic matrix-valued function $\Gamma(z)$ on CP^1 that is holomorphically invertible away from the point $a_1 = 0$ and such that the matrix U_1^0 in* (2.6) *satisfies*

$$\Gamma(z)\Gamma_4(z)U_1^0(z) = U(z)z^B, \tag{2.8}$$

where $B(z) = \operatorname{diag}((n-2)(p-1), \dots, n-2, 0)$, and the matrix $U(z)$ is holomorphic at zero.

PROOF. For an arbitrary matrix $U_1^l(z)$ we denote by ${}^l u_i^j$ the vector-valued function formed by the first j elements of the ith column of this matrix, and by ${}^l x_i^j$ the function

$${}^l x_i^j = z^{(n-2)(p-i-l)}\, {}^l u_i^j \tag{2.9}$$

We remark that the rows of the matrix $\Gamma_4(z)U_1^0$ in (2.6) have the form (2.9) with $l = 0$.

As mentioned above, the principal minors of the matrix $U_1^0(z)$ are nonzero at $a_1 = 0$; therefore, the vectors ${}^0u_1^{p-1}(0), \dots, {}^0u_{p-1}^{p-1}(0)$ are linearly independent. By Proposition 1.3.1, this implies that there are polynomials $Q_1^0(1/z), \dots, Q_{p-1}^0(1/z)$ of respective degrees $\deg Q_i^0 = (n-2)(p-i)$ such that

$$\sum_{i=1}^{p-1} Q_i^0\left(\frac{1}{z}\right) {}^0x_i^{p-1} + {}^0u_p^{p-1} = z^{(n-2)}\, {}^0w_p^{p-1}, \tag{2.10}$$

where ${}^0w_p^{p-1}(z)$ is a vector-valued function holomorphic at zero.

The matrix $\Gamma_9^1(z)\Gamma_4(z)U_1^0(z)$, where

$$\Gamma_9^1(z) = \begin{pmatrix} 1 & 0 & \cdots & 0 & 0 \\ 0 & 1 & & 0 & 0 \\ \vdots & & \ddots & \vdots & \vdots \\ 0 & & \cdots & 1 & 0 \\ Q_1^0(1/z) & & \cdots & Q_{p-1}^0(1/z) & 1 \end{pmatrix} \tag{2.11}$$

can be represented in the form

$$\Gamma_9^1(z)\Gamma_4(z)U_1^0(z) = \Gamma_4^1(z)U_1^1(z)z^{B_1}, \tag{2.12}$$

where, in turn,

$$\Gamma_4^1(z) = \operatorname{diag}(z^{(n-2)(p-2)}, \dots, z^{n-2}, 1, 1), \qquad B_1 = \operatorname{diag}(n-2, \dots, n-2, 0),$$

the matrix $U_1^1(z)$ is holomorphic at zero, and the elements of it in the principal minor of order $p-1$ coincide with the corresponding elements of $U_1^0(z)$. Therefore, the vectors ${}^1u_1^{p-2}(0), \dots, {}^1u_{p-2}^{p-2}(0)$ are linearly independent, and hence there are polynomials ${}^jQ_1^1(1/z), \dots, {}^jQ_{p-1}^1(1/z)$, $j = p-1$, p, such that

$${}^jQ_1^1\left(\frac{1}{z}\right){}^1x_1^{p-2} + \dots + {}^jQ_{p-2}^1\left(\frac{1}{z}\right){}^1x_{p-2}^{p-2} + {}^1u_j^{p-2} = z^{(n-2)}\,{}^lw_j^{p-2},$$

where the functions ${}^1w_j^{p-1}(z)$, $j = p-1$, p, are holomorphic at zero. Denote by $\Gamma_9^2(z)$ the matrix

$$\Gamma_9^2(z) = \begin{pmatrix} 1 & 0 & \cdots & 0 & 0 & 0 \\ 0 & 1 & & \vdots & \vdots & \vdots \\ \vdots & \vdots & \cdots & 1 & 0 & 0 \\ 0 & & & & & \\ {}^{p-1}Q_1^1(1/z) & {}^{p-1}Q_1^1(1/z) & \cdots & {}^{p-1}Q_{p-2}^1(1/z) & 1 & 0 \\ {}^pQ_1^1(1/z) & {}^pQ_1^1(1/z) & \cdots & {}^pQ_{p-2}^1(1/z) & 0 & 1 \end{pmatrix}.$$

The matrix $\Gamma_9^2(z)\Gamma_4^1(z)U_1^1(z)$ can be represented as

$$\Gamma_9^2(z)\Gamma_4^1(z)U_1^1(z) = \Gamma_4^2(z)U_1^2(z)z^{B_2}, \tag{2.13}$$

where

$$\Gamma_4^2(z) = \operatorname{diag}(z^{(n-2)(p-3)}, \dots, z^{(n-2)}, 1, 1, 1), \qquad B_2 = \operatorname{diag}(n-2, \dots, n-2, 0, 0),$$

the matrix $U_1^2(z)$ is holomorphic at zero, and the elements of it in the principal minor of order $p-2$ coincide with the corresponding elements of $U_1^0(z)$. Therefore, the vectors ${}^2u_1^{p-3}(0), \dots, {}^2u_{p-3}^{p-3}(0)$ are linearly independent, and so on.

Carrying out the procedure whose first two steps were described in detail above, we obtain in $p-1$ steps a matrix $\Gamma = \Gamma_9^{p-1} \cdot \ldots \cdot \Gamma_9^1$ such that $\Gamma(z)\Gamma_4(z)U_1^0(z) = U_1^{p-1}(z)z^B$, where the matrix $U_1^{p-1}(z)$ is holomorphic at zero, and $B = B_1 + \dots + B_{p-1} = \operatorname{diag}((n-2)(p-1), \dots, n-2, 0)$. We denote $U_1^{p-1}(z)$ by $U(z)$. The lemma is proved.

We pass from the space X' to $X'' = \Gamma(z)X'$, where Γ is the matrix constructed in the preceding lemma. It follows from (2.8) and Corollary 1.2.1 that the normalizations ${}''\varphi_i^j$ of this space are connected with the normalizations of the space X' by the relations

$${}''\varphi_1^j \geq {}'\varphi_1^j + (n-2)(p-j), \qquad j = 1, \dots, p. \tag{2.14}$$

We get from (2.7) and the Fuchs relation (1.31) for the original equation (1.1) that the sum Σ of the exponents ${}''\beta_i^j$ of the space X'' satisfy the inequality $\Sigma =$

$\sum_{i=1}^n \sum_{j=1}^p {}''\beta_i^j \geq 0$. We conclude from the last inequality, (1.2.21), and (1.2.22) that X'' is the solution space of a system (1.1.1) that is Fuchsian at $a_1, \dots, a_n$. The theorem is proved.

From (2.7), (2.14), (1.2.22), and Lemma 3.1.1 we get

COROLLARY 3.2.1. *The exponents* β_i^j *of the original Fuchsian equation* (1.1) *are connected with the exponents* ${}''\beta_i^j$ *of the Fuchsian system* (2.1) *constructed in the theorem by the equalities*

$$(2.15)\qquad {}''\beta_1^j = \beta_1^j - (n-2)(j-1), \qquad {}''\beta_i^j = \beta_i^j, \quad i = 2, \dots, p.$$

COROLLARY 3.2.2. *If the representation* (1.1.5) *can be realized as the monodromy representation of a Fuchsian equation* (1.1) *without additional false singular points, then it can be realized also as the monodromy representation of some Fuchsian system* (2.1).

From Theorem 3.2.1, Corollary 3.2.1, and Remark 1.4.2 we get

PROPOSITION 3.2.1. *Suppose that the representation* (1.1.5) *is reducible,* $n > 2$, $p \geq 2$, *and each of the monodromy matrices* G_i *reduces to a Jordan block. There does not exist a Fuchsian equation* (1.1) *without additional false singular points that has monodromy coinciding with* (1.1.5).

PROOF. Suppose that such an equation exists. Then by Theorem 3.2.1 and Corollary 3.2.1, there exists a Fuchsian system with the same singular points and the same monodromy such that its normalizations at a_1 are distinct. We get a contradiction to the formula (1.4.19) in Theorem 1.4.2.

We can prove Proposition 3.2.1 also directly from the Fuchs relation (1.31). Indeed, if G_i reduces to a Jordan block, then for the normalizations φ_i^j, $j = 1, \dots, p$, of equation (1.1) at a_i we have

$$(2.16)\qquad \varphi_i^1 \geq \dots \geq \varphi_i^p, \qquad i = 1, \dots, n,$$

in view of Example 1.1.3, and $\varphi_i^1, \dots, \varphi_i^m$ are the normalizations of a space Y_m, $1 \leq m < p$, that is an invariant subspace of the solution space Y of (1.1) under the action of the monodromy operators. By the substitution (1.17) we construct from Y_m a Fuchsian system with solution space X_m and with fundamental matrix T_m. This system can have additional false singular points $b_1, \dots, b_k$ at which the exponents ${}'\beta_l^j$ are positive integers, and the points themselves are regular singular points for the system constructed. By the residue theorem for the form $d \ln \det T_m(y)$ we have according to Lemma 3.1.2 that

$$\sum_{i=1}^n \sum_{j=1}^m \beta_i^j + \sum_{l=1}^k \left(\left(\sum_{j=1}^m {}'\beta_l^j \right) + c_l \right) - \frac{(n-2)m(m-1)}{2} = 0,$$

where $c_l \geq 0$, $l = 1, \dots, k$ (c_l is the order of the zero of the function $\det U_l(z)$ in the decomposition (1.2.6) for $T_m(y)|_{\tilde{O}_l}$). From the conditions ${}'\beta_l^j > 0$ and $c_l \geq 0$, $l = 1, \dots, k$, we get

$$(2.17)\qquad \sum_{i=1}^n \sum_{j=1}^m \beta_i^j \leq \frac{(n-2)m(m-1)}{2}.$$

Therefore, from (2.16),

$$\sum_{i=1}^{n}\sum_{j=1}^{p}\beta_i^j \le \frac{p}{m}\sum_{i=1}^{n}\sum_{j=1}^{m}\beta_i^j \le \frac{p}{m}\frac{(n-2)m(m-1)}{2}$$
$$= \frac{(n-2)p(m-1)}{2} < \frac{(n-2)p(p-1)}{2},$$

which contradicts the Fuchs relation (1.31).

EXAMPLE 3.2.1. For a representation (1.1.5) with arbitrary points a_1, a_2, and a_3 and with generating monodromy matrices

$$G_1 = \begin{pmatrix} 1 & c_1 \\ 0 & 1 \end{pmatrix}, \quad G_2 = \begin{pmatrix} 1 & c_2 \\ 0 & 1 \end{pmatrix}, \quad G_3 = \begin{pmatrix} 1 & -c_1-c_2 \\ 0 & 1 \end{pmatrix}, \tag{2.18}$$
$$c_1c_2(c_1+c_2) \ne 0,$$

there does not exist a Fuchsian linear differential equation of second order whose singular points are a_1, a_2, and a_3 and whose monodromy coincides with (2.16).

The proof of this follows from Proposition 3.2.1. We remark that for $p = 2$ the proof of the proposition becomes trivial. Indeed, if a Fuchsian equation with monodromy (2.18) and the three singular points exists, then the normalizations φ_i^j, $i = 1, 2, 3$, $j = 1, 2$, of this equation must satisfy the relations (2.16): $\varphi_i^1 \ge \varphi_i^2$, $i = 1, 2, 3$. Since the solution space of this equation contains the one-dimensional invariant subspace Y_1 generated by the solution u_1, the residue theorem for the form $d\ln u_1$ gives us that $\sum_{i=1}^3 \beta_i^1 \le 0$ (the function u_1 can vanish at the "additional" points $b_1, \dots, b_k$), and so $\sum_{i=1}^3(\beta_i^1+\beta_i^2) \le 0$, which contradicts the Fuchs relation $\sum_{i=1}^3(\beta_i^1+\beta_i^2) = 1$.

REMARK 3.2.1. Example 3.2.1 means, in particular, that for any values of the numbers α, β, and γ the monodromy of the hypergeometric equation

$$u'' + \frac{\gamma-(\alpha+\beta+1)z}{z(1-z)}u' - \frac{\alpha\beta}{z(1-z)}u = 0 \tag{2.19}$$

does not coincide with (2.18). It follows easily from Corollary 3.1.1 that for $n = 3$ any irreducible representation (1.1.5) of dimension $p = 2$ (satisfying the condition $\rho_0^1 = \rho_1^1 = 0$, of course) can be realized as the monodromy representation of some hypergeometric equation.

From the form of the substitution (1.17) and the character of the transformations used in the proof of Theorems 3.1.1 and 3.2.1 we get

COROLLARY 3.2.3. *The solutions of the original Fuchsian equation* (1.1) *coincide with the first components of the solutions of the system* (2.1) *constructed in Theorem* 3.2.1.

EXAMPLE 3.2.2. The constructions of Theorem 3.2.1, applied to a hypergeometric equation, lead to the well-known Fuchsian system [32]

$$df = \left(\begin{pmatrix} 0 & 0 \\ -\alpha\beta & -\gamma \end{pmatrix}\frac{dz}{z} + \begin{pmatrix} 0 & 1 \\ 0 & \gamma-(\alpha+\beta) \end{pmatrix}\frac{dz}{z-1}\right)f.$$

CHAPTER IV

Irreducible Representations

1. Technical lemmas

In what follows we need the following assertions (see [6], [7], and [29]).

LEMMA 4.1.1.([1]) *Suppose that the* $(p-l)\times l$ *matrix* $W(z)$ *is holomorphic, and the* $l\times l$ *matrix* $Y(z)$ *is holomorphically invertible in some neighborhood* O *of the point* 0. *For an arbitrary diagonal integer matrix* $C = \operatorname{diag}(c_1, \dots, c_p)$ *there is a matrix-valued function* $\Gamma(z)$ *that is meromorphic on the whole Riemann sphere, holomorphically invertible away from* 0, *and such that*

$$\Gamma(z)z^C\begin{pmatrix} Y(z)\\ W(z)\end{pmatrix} = z^{C'}\begin{pmatrix} Y(z)\\ W'(z)\end{pmatrix}, \tag{1.1}$$

where $C' = \operatorname{diag}(c_1, \dots, c_l, c'_{l+1}, \dots, c'_p)$, $c'_j > \min(c_1, \dots, c_l)$, $j > l$, *and the matrix* $W'(z)$ *is holomorphic in* O [7].

PROOF. Denote by t_l the rows of the matrix

$$z^C\begin{pmatrix} Y(z)\\ W(z)\end{pmatrix}.$$

Let $t_m = z^{c_m}w_m(z)$ be a row with index $m > l$ such that $c_m \le \min(c_1, \dots, c_l)$ and $w_m(0)\ne 0$. Since the rows $y_1(0), \dots, y_l(0)$ of the matrix $Y(0)$ are linearly independent, it follows that $w_m(0) = -\sum_{j=1}^{l} d_j y_j(0)$. Therefore, the row vector

$$\begin{aligned} t^1_m(z) &= d_1 z^{c_m - c_1} t_1(z) + \dots + d_l z^{c_m - c_l} t_l(z) + t_m(z)\\ &= z^{c_m}(d_1 y_1(z) + \dots + d_l y_l(z) + w_m(z)) \end{aligned} \tag{1.2}$$

has the form $t^1_m = z^{c^1_m} w^1_m(z)$, where either $w^1_m(z)\equiv 0$, or $w^1_m(0)\ne 0$ and $c^1_m > c_m$. If $w^1_m(z)\equiv 0$ or if $w^1_m(0)\ne 0$ and $c^1_m > \min(c_1, \dots, c_l)$, then the procedure stops here. If $c^1_m \le \min(c_1, \dots, c_l)$ and $w^1_m(0)\ne 0$, then $w^1_m(0) = -\sum_{j=1}^{l} d^1_j y_j(0)$, and we again can consider the corresponding polynomial

$$t^2_m(z) = d^1_1 z^{c^1_m - c_1} t_1(z) + \dots + d^1_l z^{c^1_m - c_l} t_l(z) + t^1_m(z),$$

and so on.

In any case we get after finitely many steps that $t^s_m(z) = z^{c'_m} w'_m(z)$, where $c'_m > \min(c_1, \dots, c_l)$ and $w'_m(z)$ is a holomorphic row vector. We consider the polynomials

$$Q^m_j = d_j z^{c_m - c_j} + d^1_j z^{c^1_m - c_j} + \dots + d^{s-1}_j z^{c^{s-1}_m - c_j}$$

([1])This lemma is an obvious generalization of Proposition 1.3.1.

in $1/z$. By construction,

$$\sum_{j=1}^{l} Q_j^m t_j(z) + t_m(z) = z^{c_m^l} w_m^l(z), \qquad m = l+1, \dots, p.$$

Let the matrix $\Gamma(z)$ in (1.1) be

$$\begin{pmatrix} 1 & 0 & \cdots & 0 & 0 & \cdots & 0 \\ 0 & 1 & \cdots & 0 & 0 & \cdots & 0 \\ \vdots & \vdots & \ddots & \vdots & \vdots & & \vdots \\ 0 & 0 & \cdots & 1 & 0 & \cdots & 0 \\ Q_1^{l+1} & Q_2^{l+1} & \cdots & Q_l^{l+1} & 1 & \cdots & 0 \\ \vdots & \vdots & & \vdots & \vdots & \ddots & \vdots \\ Q_1^p & Q_2^p & \cdots & Q_l^p & 0 & \cdots & 1 \end{pmatrix} \tag{1.3}$$

The lemma is proved.

REMARK 4.1.1. It follows from the form (1.3) of $\Gamma(z)$ that for any holomorphic $p \times m$ matrix $Z(z)$ on O the matrix $\Gamma(z)z^C Z(z)$ is also holomorphic on O.

The next assertion was proved in Lemma 3.2.4 for a special value of C.

LEMMA 4.1.2. *Suppose that the matrix $U(z)$ is holomorphically invertible in some neighborhood O of* 0, *and all the principal minors of $U(0)$ are nonzero. Then for any diagonal integer matrix $C = \operatorname{diag}(c_1, \dots, c_p)$ satisfying the condition $c_1 \geq \dots \geq c_p$ there exist a matrix $\Gamma(z)$ (which is a matrix of polynomials in $1/z$) that is holomorphically invertible away from* 0 *and a matrix $V(z)$ that is holomorphically invertible in O such that*

$$\Gamma(z)z^C U(z) = V(z)z^C. \tag{1.4}$$

PROOF. We represent the matrix $z^C U(z)$ as

$$z^C U(z) = z^{C - c_p I} U(z) z^{c_p I}, \tag{1.5}$$

where I is the identity matrix, and we apply Lemma 4.1.1 to the matrix

$$C = C - c_p I, \qquad \begin{pmatrix} (Y(z)) \\ W(z) \end{pmatrix} = \begin{pmatrix} U^{p-1}(z) \\ W_1(z) \end{pmatrix},$$

where, in turn, $U^l(z)$, $l = 1, \dots, p-1$, is formed by the intersections of the rows and columns of $U(z)$ with indices $1, \dots, l$, and $W_{p-l}(z)$ is formed by the intersections of the rows of W with indices $l+1, \dots, p$ and the columns of W with indices $1, \dots, l$.

According to Lemma 4.1.1, there exists a matrix $\Gamma_1(z)$ of the form (1.3) (with $l = p-1$) such that

$$\Gamma_1(z) z^{C - c_p I} \begin{pmatrix} U^{p-1}(z) \\ W_1(z) \end{pmatrix} = z^{C_1} \begin{pmatrix} U^{p-1}(z) \\ W_1'(z) \end{pmatrix}, \tag{1.6}$$

where $C_1 = \operatorname{diag}(c_1 - c_p, \dots, c_{p-1} - c_p, c_p^1)$, $c_p^1 > c_{p-1} - c_p$. Therefore,

$$z^{C_1} \begin{pmatrix} U^{p-1}(z) \\ W_1'(z) \end{pmatrix} = z^{C_1 - (c_{p-1} - c_p)I} \begin{pmatrix} U^{p-1}(z) \\ W_1'(z) \end{pmatrix} z^{(c_{p-1} - c_p) I_{p-1}},$$

where I_{p-1} is the $(p-1)\times(p-1)$ identity matrix. From the last formula and (1.5) we get the decomposition

$$\Gamma_1(z)z^C(U)z = z^{C_1-(c_{p-1}-c_p)I}\begin{pmatrix} U^{p-2}(z) & \Big| & Z_1 \\ W_2(z) & \Big| & \end{pmatrix} z^{D_1}, \tag{1.7}$$

where W_2 is a $2\times(p-2)$ matrix holomorphic in O, $D_1 = \operatorname{diag}(c_{p-1}, \dots, c_{p-1}, c_p)$, and Z_1 is also holomorphic in O by Remark 4.1.1. Let us apply Lemma 4.1.1 to the matrices

$$C = C_1 - (c_{p-1} - c_p)I, \qquad \begin{pmatrix} Y(z) \\ W(z) \end{pmatrix} = \begin{pmatrix} U^{p-2}(z) \\ W_2(z) \end{pmatrix}, \qquad l = p-2.$$

According to Lemma 4.1.1, there is a matrix $\Gamma_2(z)$ such that

$$\Gamma_2(z)z^{C_1-(c_{p-1}-c_p)I}\begin{pmatrix} U^{p-2}(z) \\ W_2(z) \end{pmatrix} = z^{C_2}\begin{pmatrix} U^{p-2}(z) \\ W_2'(z) \end{pmatrix}, \tag{1.8}$$

where $C_2 = \operatorname{diag}(c_1 - c_{p-1}, \dots, c_{p-2} - c_{p-1}, c'_{p-1}, c''_p)$, $c'_{p-1} > c_{p-2} - c_{p-1}$, and $c''_p > c_{p-2} - c_{p-1}$. Therefore,

$$z^{C_2}\begin{pmatrix} U^{-2}(z) \\ W_2'(z) \end{pmatrix} = z^{C_2-(c_{p-2}-c_{p-1})I}\begin{pmatrix} U^{p-2}(z) \\ W_2'(z) \end{pmatrix} z^{(c_{p-2}-c_{p-1})I_{p-2}}, \tag{1.9}$$

where I_{p-2} is the $(p-2)\times(p-2)$ identity matrix. From (1.8) and (1.9),

$$\Gamma_2\Gamma_1 z^C U(z) = z^{C_2-(c_{p-2}-e_{p-1})I}\begin{pmatrix} U^{p-3}(z) & \Big| & Z_2 \\ W_3(z) & \Big| & \end{pmatrix} z^{D_2},$$

where W_3 is a $3\times(p-3)$ matrix, W_3 and Z_2 are holomorphic in O, and $D_2 = \operatorname{diag}(c_{p-2}, \dots, c_{p-2}, c_{p-1}, c_p)$. And so on.

As a result we get in at most $p-1$ steps (the first two of which were described above) a matrix $\Gamma(z) = \Gamma_{p-1}\cdots\Gamma_1$ such that the decomposition (1.4) holds for some matrix $V(z)$ holomorphic in O.

Since

$$\det V(0) = \lim_{z\to 0}\det\Gamma(z)\det U(0) = \det U(0) \neq 0,$$

we get that $V(z)$ is holomorphically invertible at 0. (Here we use the form (1.3) of the matrix Γ_i, which implies that $\det\Gamma_i(z)\equiv 1$.)

COROLLARY 4.1.1. *The matrix* $V(z)$ *in* (1.4) *has the following properties*:

a) $V(0)$ *is upper triangular*;

b) $V(z)z^C = z^C V^0(z)$, *where* $V^0(z)$ *is some matrix that is holomorphically invertible at zero.*

The proof follows immediately from the constructions of the preceding lemma.

LEMMA 4.1.3. *For any matrix* $U(z)$ *holomorphically invertible in some neighborhood* O *of* 0 *and for any diagonal integer matrix* $C = \operatorname{diag}(c_1, \dots, c_p)$ *there exist a matrix* $\Gamma(z)$ *(which is a polynomial in* $1/z$*) holomorphically invertible away from zero and a matrix* $V(z)$ *holomorphically invertible in* O *such that*

$$\Gamma(z)z^C U(z) = V(z)z^D, \tag{1.10}$$

where the matrix $D = \operatorname{diag}(d_1, \dots, d_p)$ *is a permutation of the diagonal elements of* C.

If for some other matrix $\Gamma(z)$ *with the same properties* (1.10) *holds with some diagonal integer matrix* D, *then*

$$c^1 \le d^1, \qquad c^p \ge d^p, \tag{1.11}$$

where

$$c^1 = \max(c_1, \dots, c_p), \qquad d^1 = \max(d_1, \dots, d_p),$$
$$c^p = \min(c_1, \dots, c_p), \qquad d^p = \min(d_1, \dots, d_p).$$

PROOF. We consider first the case when the diagonal elements of C are arranged in nonincreasing order. With the help of a nonsingular constant matrix S we permute the columns of $U(z)$ so that all the principal minors of $U' = US$ are nonzero. Applying Lemma 4.1.2 to U', we get

$$\Gamma(z)z^C U'(z) = V'(z)z^C,$$

and hence

$$\Gamma(z)z^C U(z) = \Gamma(z)z^C U'(z)S^{-1} = V'(z)z^C S^{-1} = V'(z)S^{-1}z^{SCS^{-1}} = V(z)z^D.$$

If the elements $c_1, \dots, c_p$ are not ordered, then there is a constant matrix S' such that $(S')^{-1}CS' = C'$, where $C' = \operatorname{diag}(c'_1, \dots, c'_p)$, and the elements $c'_1, \dots, c'_p$ are arranged in nonincreasing order. For the matrix $z^{C'}(S')^{-1}U(z)$ we consider the corresponding matrix $\Gamma'(z)$. In this case the matrix $\Gamma'(S')^{-1}$ should be taken as Γ for the matrices C and $U(z)$ in (1.10).

We prove the second assertion of the lemma. Assume that $d^p > c^p$ for some matrices $\Gamma(z)$ and D in (1.10). Let $d^p = d_l$ and $c^p = c_m$ for suitable l and m. From (1.10),

$$\Gamma(z) = V(z)z^D U^{-1}(z)z^{-C}. \tag{1.12}$$

The element u'_{sm} of the matrix $z^D U^{-1} z^{-C}$ has the form $u'_{sm} = u_{sm} z^{d_s - c_m}$, where u_{sm} is a holomorphic function. Therefore, the mth column γ_m of $\Gamma(z)$ has the form $\gamma_m = v_m z^{d_l - c_m}$ for some holomorphic column $v_m(z)$. Since $d^p > c^p$ by assumption, the components of the column γ_m are holomorphic at zero, and $\gamma_m(0) = 0$. Since γ_m is holomorphic away from zero, Liouville's theorem gives us that $\gamma_m(z) \equiv 0$, which contradicts the holomorphic invertibility of $\Gamma(z)$ away from 0.

The first inequality in (1.11) is proved similarly. In place of (1.12) it is necessary to consider the formula $\Gamma^{-1}(z) = z^C U(z) z^{-D} V^{-1}(z)$ and in place of γ_m the column γ^t of $\Gamma^{-1}(z)$, where $d^1 = d_t$ for a suitable t.

2. The maximal Fuchsian weight of a representation

The notation of §4 in Chapter I will be used in this section.

Consider the bundle $\widetilde{F} = (\widetilde{F}_E, CP^1 \setminus \mathscr{D}, \widetilde{\pi}_F)$ constructed from a monodromy representation (1.1.5), and denote by $\widetilde{\nabla}$ a holomorphic connection in this bundle (see §6 of Chapter II). As in the two-dimensional case studied in §6 of Chapter II, we consider all possible extensions $(\widetilde{F}, \widetilde{\nabla})$ to the whole of CP^1 for which the connection has at most logarithmic singularities at the points of $\mathscr{D}$. It follows from Theorem 1.4.1 that all such extensions $(F^\varphi, \widetilde{\nabla})$ coincide with the extensions constructed from all possible admissible tuples φ (see Definition 1.4.1).

We consider the decomposition (1.4.14) for the bundle F^φ along with the weight $\gamma(\varphi)$ of this bundle (see Definition 1.4.2).

DEFINITION 4.2.1. The number

$$\gamma_m(\chi) = \sup_\varphi \gamma(\varphi)$$

is called the *maximal Fuchsian weight* of the representation χ. (Here the supremum is taken over all admissible tuples φ and over all sets of S; see Remark 1.4.1)

It is interesting that the irreducible representations (1.1.5) are distinguished among all others by the finiteness of the maximal Fuchsian weight.

THEOREM 4.2.1. *A representation* (1.1.5) *is irreducible if and only if its maximal Fuchsian weight is finite. A sharp estimate for the maximal Fuchsian weight of an irreducible representation is*

$$\gamma_m(\chi) \leq (n-2)p(p-1)/2. \tag{2.1}$$

PROOF. *Necessity.* Let χ be an irreducible representation, and φ an admissible tuple. We consider the bundle F^φ and the decomposition (1.4.14) for this bundle. Proceeding as in Lemma 1.4.2, we construct a system with Fuchsian singular points $a_2, \dots, a_n$ and with a regular singular point a_1 such that a fundamental matrix $T_1(y)$ of this system in $\widetilde{O}_1$ has the form

$$T_1(y) = (z-a_1)^{C^\varphi} V_1'(z)(z-a_1)A^{\varphi_1}(y-a_1)^{E_1^0} \tag{2.2}$$

(see the formula preceding (1.4.15), with $T = T_1 S^{-1}$, $V_1' = \widetilde{V}^{-1}S$, $y^* = y \in \widetilde{O}_1$, and $\chi(\sigma) = I$), where the diagonal integer matrix C^φ determines the splitting type (1.4.14) of F^φ.

We apply to $T_1(y)$ the construction of Theorem 3.1.1. It follows from Remark 3.1.2 that in this case the inequality (3.1.29) holds with $\sum_{i=1}^p (k_1 - k_i)$ replaced by $\gamma(\varphi) = \sum_{i=1}^p (c_1^\varphi - c_i^\varphi)$:

$$m \leq \frac{(n-2)p(p-1)}{2} - \left(\gamma(\varphi) + l - 1 + \sum_{i=1}^n s_i\right). \tag{2.3}$$

Since $m \geq 0$, $l \geq 1$, and $s_i \geq 0$, the last inequality gives us that

$$\gamma(\varphi) \leq (n-2)(p(p-1)/2,$$

which by the arbitrariness of φ implies (2.1).

Sufficiency. Assume that the representation (1.1.5) is reducible. It can be assumed that the monodromy matrices $\chi(\sigma_i)$ have the form([2])

$$\chi(\sigma_i) = \begin{pmatrix} G_i^1 & * \\ 0 & G_i^2 \end{pmatrix}, \qquad i = 1, \dots, n. \tag{2.4}$$

We consider a system (1.1.1) that is Fuchsian at $a_2, \dots, a_n$, regular at a_1, and such that all its normalizations at $a_2, \dots, a_n$ are equal to zero. (The existence of

([2])In Chapters IV and V the generators of the group $\pi_1(CP^1 \setminus \mathscr{D}, z_0)$ are denoted by σ_i instead of g_i (see §1 of Chapter I after the formula (1.1.5)).

such a system was proved in Proposition 1.4.1.) Moreover, suppose that the matrix form ω of this system also has the appearance (2.4) (with G_i^j replaced by ω^j). (The existence of such a system for arbitrary p will be proved in Theorem 5.1.1; for $p = 2$ and 3 this follows from Lemma 1.3.8.) Denote by $T(y)$ a fundamental matrix of this system of the form (2.4) (with G_i^j replaced by T_1^j). Then in the decomposition (1.1.10)

$$T(y) = M_1(z)(y-a_1)^{E_1}.$$

For $T(y)$ the meromorphic matrix-valued function $M_1(z)$ has the form

$$M_1(z) = \begin{pmatrix} M_1^1 & * \\ 0 & M_1^2 \end{pmatrix}. \tag{2.5}$$

Suppose that the maximal order of a pole for elements of M_1^1 is equal to p_1, and the maximal order for elements of M_1^2 is p_2. Applying Lemma 1.3.9 to the matrices $(z-a_1)^{p_1}M_1^1$ and $(z-a_1)^{p_2}M_1^2$, we get matrices Γ_1, C_1, V_1 and Γ_2, C_2, V_2 (see (1.3.26)) such that $\Gamma_i(z)(z-a_1)^{p_i}M_1^i = (z-a_1)^{C_i}V_i$, $i = 1, 2$.

We let $\Gamma = \operatorname{diag}(\Gamma_1, \Gamma_2)$ and pass from the system with fundamental matrix $T(y)$ to a system with fundamental matrix $T'(y) = \Gamma(z)T(y)$. We have

$$T'(y) = (z-a_1)^C V(z)(z-a_1)^{E_1}, \tag{2.6}$$

where

$$C = \operatorname{diag}((z-a_1)^{-p_1}C_1, (z-a_1)^{-p_2}C_2), \qquad V(z) = \begin{pmatrix} V_1 & W \\ 0 & V_2 \end{pmatrix},$$

V_1 and V_2 are holomorphically invertible at a_1, and W has in general a pole of some order p_3 at a_1.

Suppose that the blocks G_i^1 in (2.4) are $l \times l$ matrices. We consider the following extension F^φ of the bundle $\widetilde{F}$ constructed from the given representation (1.1.5):

$$\varphi^2 = \cdots = \varphi^n = 0, \qquad \varphi_1^1 = \cdots = \varphi_1^l = d > 0, \qquad \varphi_1^j = -p_3, \quad j > l. \tag{2.7}$$

Denote by K the matrix $\operatorname{diag}(\varphi_1^1, \dots, \varphi_1^p)$. Then

$$V(z)(z-a_1)^{-K} = (z-a_1)^{-K}V_0(z),$$

where $V_0(z)$ is holomorphically invertible at a_1. The tuple φ is admissible, and by (2.6)

$$T'(y) = (z-a_1)^{C-K}V_0(z)(z-a_1)^K(y-a_1)^{E_1}$$

for $y \in \widetilde{O}_1$, and

$$T'(y) = V_i(z)(y-a_i)^{E_i}$$

for $y \in \widetilde{O}_i$ by construction. Proceeding as in the proof of Theorem 1.4.1, we get that the diagonal elements of $C - K$ (taken in nonincreasing order) determine the splitting type of the bundle F^φ. Since C and p_3 are fixed, for any previously specified number D there is a number $d > 0$ such that $\operatorname{tr}(C-K) < p(c_p + p_3) - D$. It follows from $\max_i(c_i - k_i) \geq c_p + p_3$ that

$$\gamma(\varphi) \geq p(c_p + p_3) - \operatorname{tr}(C-K) > D.$$

Therefore, $\gamma_m(\chi) = +\infty$. The sufficiency is proved (modulo Theorem 5.1.1).

The fact that the estimate (2.1) is sharp will be proved in §4.

3. A positive solution for an irreducible representation

THEOREM 4.3.1. *The Riemann-Hilbert problem has a positive solution for any irreducible representation* χ [7].

PROOF. Proceeding as in Theorem 4.2.1, we construct from the given representation χ a system with Fuchsian singular points $a_2, \dots, a_n$ and a regular singular point a_1 such that a fundamental matrix $T_1(y)$ of this system on $\widetilde{O}_1$ has the form (4.2.2):

$$T_1(y) - (z-a_1)^{C^\varphi} V_1'(z)(z-a_1)^{A\varphi_1}(y-a_1)^{E_1^0}, \tag{3.1}$$

where $\varphi = (\varphi_1, \dots, \varphi_p)$ is an admissible tuple satisfying the condition

$$\varphi_1^i - \varphi_1^{i+1} > \frac{(n-2)p(p-1)^2}{2}, \qquad i = 1, \dots, p-1, \tag{3.2}$$

for $\varphi_1 = (\varphi_1^1, \dots, \varphi_1^p)$, $C^\varphi = \operatorname{diag}(c_1^\varphi, \dots, c_p^\varphi)$, and $c_1^\varphi \geq \cdots \geq c_p^\varphi$. According to Theorem 4.2.1,

$$c_1^\varphi - c_p^\varphi \leq \sum_{i=1}^{p-1}(c_i^\varphi - c_{i+1}^\varphi) \leq \sum_{j=1}^{p}(c_1^\varphi - c_j^\varphi) \leq \frac{(n-2)p(p-1)}{2}. \tag{3.3}$$

We apply Lemma 4.1.3 to the matrices $V_1'(z)$ and C^φ. By this lemma, there is a meromorphic matrix $\Gamma(z)$ on the whole Riemann sphere that is holomorphically invertible away from a_1 and such that

$$\Gamma(z)(z-a_1)^{C^\varphi} V_1'(z) = V(z)(z-a_1)^D, \tag{3.4}$$

where the diagonal matrix D is obtained from C^φ by a permutation of the diagonal elements.

We pass from the system with fundamental matrix $T_1(y)$ to a system with matrix $T_1'(y) = \Gamma(z)T_1(y)$. From (3.1) and (3.4),

$$T_1'(y) = V(z)(z-a_1)^{D+A\varphi_1}(y-a_1)^{E_1^0}, \tag{3.5}$$

where $V(z)$ is holomorphically invertible, and in view of (3.2) and (3.3) $D + A\varphi_1$ is a diagonal integer matrix with nonincreasing diagonal elements. Since E_1^0 is an upper triangular matrix, the matrix $(z-a_1)^{D+A\varphi_1}E_1^0(z-a_1)^{-D-A\varphi_1}$ is holomorphic at a_1. Therefore, we get from Corollary 1.2.2 that the system constructed is Fuchsian at a_1. The theorem is proved.

We have actually proved somewhat more than we wanted. We have obtained a Fuchsian system with preassigned normalizations at the points $a_2, \dots, a_n$. What is more, we have proved the following assertion.

THEOREM 4.3.2. *For any irreducible representation χ and any integer $d > 0$ there exists a Fuchsian system with monodromy χ and with any preassigned admissible normalizations at the points $a_2, \dots, a_n$ such that its normalizations φ_1^j at a_1 satisfy the inequalities*

$$\varphi_1^j - \varphi_1^{j+1} > d, \qquad j = 1, \dots, p-1. \tag{3.6}$$

4. An exact expression for the number of additional false singular points

In this section we give the exact value for the minimal number of additional false singular points arising in the construction of a scalar Fuchsian equation with given irreducible monodromy (see §1 in Chapter III). Recall that if $u_1, \dots, u_p$ is a basis in the space of solutions of the Fuchsian equation (3.1.1) with given monodromy (1.1.5), then this equation can be written in the form

$$\frac{1}{W(y)}\begin{vmatrix} u & u_1 & \cdots & u_p \\ \frac{du}{dz} & \frac{du_1}{dz} & \cdots & \frac{du_p}{dz} \\ \vdots & \vdots & \cdots & \vdots \\ \frac{d^p u}{dz^p} & \frac{d^p u_1}{dz^p} & \cdots & \frac{d^p u_p}{dz^p} \end{vmatrix} = 0, \tag{4.1}$$

where $W(y)$ is the Wronskian of the functions $u_1, \dots, u_p$. The additional false singular points of this equation are the points $b_1, \dots, b_t$ that are the projections of the zeros of $W(y)$. Denote by p_j the order of the zero of $W(y)$ at (any) one of the points in $\pi^{-1}b_j$.

Denote by m the number of additional singular points (4.1) with multiplicity counted, that is, $m = p_1 + \cdots + p_t$, and by m^0 the number

$$m^0 = \min_{\mathrm{FE}} m,$$

where the minimum is taken over all Fuchsian equations with the given monodromy.

THEOREM 4.4.1. *The minimal possible number* m_0 *of additional false singular points of a scalar Fuchsian equation* (3.1.1) *with given irreducible monodromy* (1.1.5) *is*

$$m^0 = \frac{(n-2)p(p-1)}{2} - \gamma_m(\chi), \tag{4.2}$$

where $\gamma_m(\chi)$ *is the maximal Fuchsian weight of the representation* χ *(see Definition* 4.2.1).

PROOF. It can be assumed without loss of generality that $a_1 = 0$ (this can always be achieved by a suitable linear fractional transformation of the Riemann sphere).

We consider a basis $u_1, \dots, u_p$ in the solution space of a Fuchsian equation with minimal possible number m^0 of additional false singular points $b_1, \dots, b_t$. It follows from the condition of minimality that for any $j = 1, \dots, t$

$$u_i(\pi^{-1}b_j) \neq 0 \tag{4.3}$$

for some i. Indeed, if for some j all the u_i were zero in $\pi^{-1}b_j$, then the Fuchsian equation with solutions $v(y) = (v/(z-b_j))u(y)$ would have fewer than m^0 additional false singular points (counting multiplicity), which is impossible.

From $u_1, \dots, u_p$ we construct a system of equations on the Riemann sphere with unknown function $f(z) = (f^1, \dots, f^p)^T$ (where T denotes transposition) as follows:

$$f^j = \left(\prod_{i=1}^{n}(z-a_i)^{j-1}\right)\left(\prod_{r=1}^{t}(z-b_r)^{j-1}\right)\frac{d^{j-1}u}{dz^{j-1}}. \tag{4.4}$$

Lemma 3.1.1 gives us that this system is Fuchsian at the points $a_2, \dots, a_n$ and $b_1, \dots, b_j$. It follows from (1.2.6) and (4.4) that a fundamental matrix $T_1(y)$ of this system has in $\widetilde{O}_1$ the form([3])

$$T_1(y) = U_1(z) z^{\Lambda_1} y^{E_1}, \tag{4.5}$$

where U_1 is holomorphically invertible at zero.

The system constructed is only regular at ∞. Proceeding just as in Theorem 3.2.1 (Lemmas 3.2.1–3.2.3; the only difference is an increase in the number of singular points because of $b_1, \dots, b_t$), we can show that there exists a meromorphic lower triangular matrix Γ' on the Riemann sphere that is holomorphically invertible away from 0 and ∞ and such that the system with fundamental matrix $Y(y) = \Gamma' T_1(y)$ is holomorphic at ∞, and $Y(y)$ has in $\widetilde{O}_1$ the decomposition (analogous to (2.2.6))

$$Y(y) = z^{K'} U'(z) z^{\Lambda} y^{E_1}, \tag{4.6}$$

where $K' = \operatorname{diag}((n+t-2)(p-1), (n+t-2)(p-2), \dots, 0)$, and

$$\Lambda = \Lambda_1 - (n+t-2)(p-1)I.$$

We consider the decomposition (1.2.6) for the constructed system at the point b_r:

$$Y_r(z) = \widetilde{U}_r(z)(z-b_r)^{\widetilde{\Lambda}_r}. \tag{4.7}$$

Note that $\widetilde{\varphi}_r^p = 0$ in view of (4.3) (recall that $\widetilde{\Lambda}_r = \operatorname{diag}(\widetilde{\varphi}_r^1, \dots, \widetilde{\varphi}_r^p)$).

Suppose that the normalizations $\widetilde{\varphi}_r^j$ are equal to zero for $j > l$ and $\widetilde{\varphi}_r^l > 0$. Just as in Lemma 3.2.1 it can be shown that all the minors of $\widetilde{U}_r(b_r)$ obtained by intersecting the first l rows and the last l columns are nonzero at b_r (since the local monodromy at this point is lacking). Therefore, by Lemma 4.1.1, there exists a nonsingular constant lower triangular matrix S_l of the form

$$S_l = \begin{pmatrix} I^l & 0 \\ 0 & S_l' \end{pmatrix}$$

such that the elements of $S_l Y_r(y)$ with indices (k, m), $k > l$, $m > p - l$, are equal to zero at b_r. Therefore,

$$\left(\frac{z}{z-b_r}\right)^D S_l Y_r(y) = \widetilde{U}_r'(z)(z-b_r)^{\widetilde{\Lambda}_r + D - I}, \tag{4.8}$$

where $D = \operatorname{diag}(0, \dots, 0, 1, \dots, 1)$ (with zeros in the first l places), and $\widetilde{U}_r'(z)$ is holomorphically invertible at b_r.

We transform our system as follows:

$$Y' = \left(\frac{z}{z-b_r}\right)^D S_l Y. \tag{4.9}$$

It follows from the form of the matrices S_l and D that this transformation does not change the first l rows of Y and decreases by 1 each of the first l normalizations at b_r. Therefore, if the first $m < l$ normalizations of the new system at b_r are nonzero, then we can again use the procedure described above, and so on.

([3])Here and below the matrices of normalizations will be denoted by Λ_i instead of A_i.

Applying the above procedure to all the points $b_1, \dots, b_t$, we get in finitely many steps a system that is Fuchsian at $a_2, \dots, a_n$, regular at zero, and holomorphic at the remaining points. A fundamental matrix $Y''(y)$ of this system at zero has the form

$$Y''(y) = U''(z)z^{\Lambda}y^{E_1}, \tag{4.10}$$

where the matrix U'' is holomorphic at zero, and, in turn, its first row has the form

$$z^{(n-2+t)(p-1)} \cdot \text{(a holomorphic row vector)}$$

in view of (4.6). We note that

$$\det U'' = z^{\operatorname{tr} K'} + \sum_{r=1}^{t} \operatorname{tr} \widetilde{\Lambda}_r. \tag{4.11}$$

The fact that the sum of the orders of the zeros of $W(y)$ over all $b_1, \dots, b_t$ is equal to m^0 and the form of the transformations (4.4) and (4.7) give us that

$$\sum_{r=1}^{t} \operatorname{tr} \widetilde{\Lambda}_r = m^0 + \frac{tp(p-1)}{2},$$

since $W(y) = \det Y(y)(z - b_r)^{p(p-1)/2}$ at the point $\pi^{-1}(b_r)$. Therefore,

$$\det U'' = z^{(n-2+2t)p(p-1)/2+m^0}. \tag{4.12}$$

According to Lemma 1.3.9, there exists a matrix $\Gamma(z)$ that is holomorphically invertible away from zero and such that

$$Z(y) = \Gamma(z)Y''(y) = z^K U^0(z) z^{\Lambda} y^{E_1}, \tag{4.13}$$

where $K = \operatorname{diag}(k_1, \dots, k_p)$, $k_1 \geq \cdots \geq k_p \geq 0$, and U^0 is holomorphically invertible at zero. Therefore, $z^{\operatorname{tr} K} = \det U''$.

We show that

$$k_1 \geq (n-2+t)(p-1). \tag{4.14}$$

Indeed, otherwise the first row of the matrix

$$\Gamma^{-1}(z) = Y''Z^{-1} = U''(U^0)^{-1}z^{-K}$$

would be holomorphic at zero, and would vanish there. (Since this row has the form $z^{((n-2+t)(p-1)-k_1)} \cdot$ (a holomorphic row vector).) The holomorphic invertibility of Γ at zero would give us that this row is identically equal to zero, which is impossible.

The system constructed determines a meromorphic trivialization (which is holomorphic away from zero) of the vector bundle F^{φ}, where for $i > 1$ the tuple of φ_i^j coincides with the corresponding normalizations of the original Fuchsian equation, while the tuple of φ_1^j coincides with the diagonal elements of the matrix Λ, and the matrix K determines the splitting type (1.4.14) of this bundle (see §4 of Chapter I). It follows from (4.12) and (4.14) that the weight γ' of this bundle satisfies the inequality

$$
\begin{aligned}
\gamma' &= \sum_{i=1}^{p}(k_1 - k_i) = pk_1 - \operatorname{tr} K = pk_1 - \left(\frac{(n-2+2t)p(p-1)}{2} + m^0\right) \\
&\geq (n-2+t)p(p-1) - \left(\frac{(n-2+2t)p(p-1)}{2} + m^0\right) \\
&= \frac{(n-2)p(p-1)}{2} - m^0 .
\end{aligned}
\tag{4.15}
$$

Therefore,

$$\gamma_m(\chi) + m^0 \geq \gamma' + m^0 \geq (n-2)p(p-1)/2.$$

But it follows from (4.2.3) that

$$\gamma_m(\chi) + m^0 \leq (n-2)p(p-1)/2.$$

The assertion of the theorem now follows from the last two inequalities.

COROLLARY 4.4.1. *An irreducible representation* (1.1.5) *can be realized as the monodromy representation of some scalar Fuchsian differential equation without additional singular points if and only if the maximal Fuchsian weight of this representation takes its maximal possible value* $(n-2)p(p-1)/2$.

As shown in §2 of Chapter III, each scalar Fuchsian equation on the Riemann sphere can be transformed into a Fuchsian system with the same singular points and monodromy. The corollary above answers the question of how to describe all the Fuchsian systems that can be obtained in this way. In particular, it implies that the system in Example 2.2.1 is such a system.

5. Analytic transformation to standard Birkhoff form

We consider a system (1.1.1) of linear differential equations written as

$$z\frac{df}{dz} = A(z)f, \tag{5.1}$$

where $A(z)$ is a $p\times p$ matrix-valued function of the complex variable z of the form

$$A(z) = z^r\sum_{n=0}^{\infty} A_n z^{-n}, \qquad A_0 \neq 0, \ r \geq 0,$$

where the matrix series converges in some neighborhood of ∞.

The transformation

$$u = \Gamma(z)f \tag{5.2}$$

carries the system (5.1) into the system

$$z\frac{du}{dz} = B(z)u, \tag{5.3}$$

where

$$B(z) = z\frac{d\Gamma}{dz}\Gamma^{-1} + \Gamma A(z)\Gamma^{-1}.$$

The transformation (5.2) is said to be analytic if the matrix $\Gamma(z)$ is holomorphically invertible in some neighborhood of ∞.

If the matrix $B(z)$ in (5.3) is an rth-degree polynomial in z, then (5.3) is called the *Birkhoff standard form* for equation (5.1).

Birkhoff [37] thought that any system (5.1) could be reduced by an analytic transformation to Birkhoff standard form, but Gantmakher [11] constructed a counterexample to this assertion. It turned out that Birkhoff's proof works only for the case when the monodromy matrix of (5.1) is diagonalizable.

We say that a system (5.1) is reducible if there is a matrix $\Gamma(z)$ that is holomorphically invertible in a neighborhood of ∞ and such that the transformation (5.2) reduces (5.1) to a system (5.3) with lower triangular block matrix

$$B(z) = \begin{pmatrix} B_1 & 0 \\ * & B_2 \end{pmatrix} \tag{5.4}$$

It was proved by Jurkat, Lutz, and Peyerimhoff for $p = 2$ [44] and by Balser [32] for $p = 3$ that each irreducible system (5.1) ("generic system" in the terminology of [32]) can be reduced by an analytic transformation to Birkhoff standard form. Here we prove that an analogous result holds for arbitrary p.

THEOREM 4.5.1 ([34], [35]). *Any irreducible system* (5.1) *can be reduced by an analytic transformation to Birkhoff standard form.*

PROOF. We consider a fundamental matrix $T(y)$ of (5.1) in the form (1.1.10):

$$T(y) = M(z)z^E,$$

where $M(z)$ is a single-valued matrix function with nonvanishing determinant $\det M(z)$ in some deleted neighborhood K of ∞, and E is reduced to Jordan normal form.

We consider an arbitrary diagonal integer matrix $\Lambda = \operatorname{diag}(\lambda_1, \dots, \lambda_p)$ such that

$$\lambda_1 \geq \cdots \geq \lambda_p. \tag{5.5}$$

The matrix $M(z)z^{-\Lambda}$ can be regarded as the transition function of some vector bundle on the Riemann sphere CP^1 with coordinate neighborhoods $K \cup \{\infty\}$ and $CP^1 \setminus \{\infty\}$. According to the Birkhoff–Grothendieck theorem (1.4.14), there exist a matrix $\Gamma(z)$ holomorphically invertible in a neighborhood of ∞ and a matrix $U(z)$ holomorphically invertible in the complex plane such that

$$\Gamma(z)M(z)z^{-\Lambda} = z^C U(z), \tag{5.6}$$

where $C = \operatorname{diag}(c_1, \dots, c_p)$, $c_i \in \mathbb{Z}$, and $c_1 \geq \cdots \geq c_p$.

For an arbitrary system (5.1) we consider an arbitrary diagonal integer matrix Λ satisfying (5.5) together with the corresponding matrix C in the decomposition (5.6).

PROPOSITION 4.5.1. *If the system* (5.1) *is irreducible, then the elements* c_i *of the matrix* C *satisfy the inequalities*

$$c_i - c_{i+1} \leq r, \qquad i = 1, \dots, p-1. \tag{5.7}$$

PROOF. Assume the contrary. Suppose that for some index $k = 1, \dots, p-1$

$$c_k - c_{k+1} > r. \tag{5.8}$$

We consider a system with fundamental matrix

$$T'(y) = z^C U(z) z^\Lambda y^E. \tag{5.9}$$

The only singular points of this system on the Riemann sphere are the points 0 and ∞ (since $U(z)$ is holomorphically invertible everywhere except at ∞), and the coefficient matrix $A' = z\frac{dT'}{dz}(T')^{-1}$ of this system has a pole of order r at ∞. The last assertion follows from the fact that this system is obtained from the original system by means of a transformation $\Gamma(z)$ that is analytic in some neighborhood of ∞.

On the other hand, it follows from (5.9) that

$$A'(z) = C + z^C\left[\frac{dU}{dz}U^{-1} + U(\Lambda + L)U^{-1}\right]z^{-C}, \tag{5.10}$$

where $L = z^\Lambda E z^{-\Lambda}$. The inequality (5.5) and the fact that E has upper triangular form give us that L is holomorphic on the whole complex plane. Therefore, the matrix in square brackets in (5.10) is holomorphic everywhere except at ∞. Denote this matrix by $W(z)$. Then

$$A'(z) = C + z^C W(z) z^{-C}.$$

Since an element a'_{ij} of A' is connected with an element w_{ij} of W by the relation

$$a'_{ij} = z^{c_i - c_j} w_{ij}, \qquad i \neq j,$$

we get from the assumption (5.8) that for $i = 1, \dots, k$ and $j = k+1, \dots, p$

$$c_i - c_j > r.$$

Therefore, for each such pair i, j of indices the element a'_{ij} has a zero of order $m > r$ at 0, while the order of the pole of this element at ∞ does not exceed r. Hence $a'_{ij} \equiv 0$, $i = 1, \dots, k$, $j = k+1, \dots, p$; that is, the original system (5.1) is reducible. But this contradicts the condition of the proposition. The contradiction means that the inequality (5.7) does hold.

Let us continue the proof of Theorem 4.5.1. We consider a matrix Λ satisfying the inequalities (5.5) and such that

$$\lambda_i - \lambda_{i+1} > r(p-1), \qquad i = 1, \dots, p-1. \tag{5.11}$$

We consider the corresponding matrices Γ and C in (5.6). According to Lemma 4.1.3, applied to the matrix $z^C U(z)$ in (5.9), there is a matrix $\Gamma_1(z)$ that is holomorphically invertible away from ∞ for which the decomposition (4.1.10) holds.

Under the action of the analytic (in some neighborhood of ∞) transformation

$$u = \Gamma_1(z)\Gamma(z) f$$

the original system (5.1) passes into a system (5.3) with fundamental matrix

$$T''(y) = \Gamma_1(z)\Gamma(z)T(y) = V(z) z^{D+\Lambda} y^E,$$

where the matrix $V(z)$ is holomorphically invertible on the whole Riemann sphere except at ∞.

It follows from Lemma 4.1.3 and Proposition 4.5.1 that the elements d_j of the diagonal integral matrix D satisfy the inequalities

$$d_j - d_{j+1} \leq r(p-1), \qquad j = 1, \dots, p-1. \tag{5.12}$$

Indeed,

$$|d_j - d_{j+1}| \le \max_j d_j - \min_i d_i = c_1 - c_p$$
$$= (c_1 - c_2) + (c_2 - c_3) + \cdots + (c_{p-1} - c_p) \le r(p-1),$$

since D is a permutation of the diagonal elements of C.

Thus, it follows from (5.12) and (5.11) that the diagonal elements of $D+\Lambda$ are arranged in nonincreasing order. Since E has upper triangular form, we again get that the elements of the matrix

$$L' = z^{D+\Lambda} E z^{-D-\Lambda}$$

are entire functions. Therefore, the elements of the matrix

$$B(z) = z\frac{dT''}{dz}(T'')^{-1} = z\frac{dV}{dz}V^{-1} + V(D+\Lambda+L')V^{-1}$$

are also entire functions. By the holomorphic invertibility of Γ_1 and Γ at ∞, the matrix $B(z)$ has a pole of order r there, and hence is an rth degree polynomial in z. The theorem is proved.

REMARK 4.5.1. The proof of Theorem 4.5.1 is similar to the main part of the proof of Theorem 4.3.1 on realization of any irreducible representation by a Fuchsian system. In both cases the proof is based on the result that the maximal weight of a certain family of bundles on the Riemann sphere is finite.

6. Connections with logarithmic singularities and the Fuchsian weight

We consider a Fuchsian system

$$\frac{df}{dz} = \left(\sum_{i=1}^{n} \frac{B_i}{z-a_i}\right) f, \tag{6.1}$$

its monodromy (1.1.5), and the canonical extension F^0 of the vector bundle $\widetilde{F}$ (see §4 of Chapter I). The splitting type of the bundle F^0 can be algorithmically computed with the help of the system (6.1) as follows.

For the matrix B_1 in (6.1), with normalizations $\varphi_1^1, \dots, \varphi_1^p$ and with exponents $\beta_1^1, \dots, \beta_1^p$, suppose that $\varphi_1^1 \ne 0$. It can be assumed without loss of generality that $\varphi_1^1 > 0$. (The case $\varphi_1^1 < 0$ is analized similarly.)

We consider a gauge transformation

$$g_1 = \Gamma_1 f, \tag{6.2}$$

where Γ_1 is a nonsingular constant matrix such that the coefficient matrix $B_1' = \Gamma_1 B_1 \Gamma_1^{-1}$ of the new system has normal Jordan form with the first block, corresponding to the eigenvalue φ_1^1, of dimension p_1.

We also consider the transformation

$$g_2 = (x-a_1)^D g_1 \tag{6.3}$$

with

$$D = \operatorname{diag}(\underbrace{-1, \dots, -1}_{p_1}, 0, \dots, 0).$$

Under the action of (6.2) and (6.3) the original system is transformed into a system with coefficient matrices B_i'' that is Fuchsian at the points $a_1, \ldots, a_n$ with the additional false singular point ∞ and is such that the eigenvalues of B_1'' at a_1 are equal to $\beta_1^1 - 1, \ldots, \beta_1^{p_1} - 1, \beta_1^{p_1+1}, \ldots, \beta_1^p$ (see [16] or [26], Lemma 11.3). Thus, the values of p_1 of the normalizations of the original system are decreased by 1.

By implementing sufficiently many times the procedure whose first step was just described, we can get a system (1.1.1) that is Fuchsian at $a_1, \ldots, a_n$, has an additional false singular point at ∞, and is such that all the normalizations of this system at a_1 are equal to zero.

Proceeding similarly, we can transform (6.1) into a system with zero normalizations at all the points $a_1, \ldots, a_n$. This system will be Fuchsian at these points and will have a unique additional false singular point at ∞. The corresponding matrix $T(z)$ of the transformation will be meromorphic at ∞ and holomorphically invertibility away from ∞. This matrix can be algorithmically computed from the coefficient matrices of the original system.

The matrix $T(z)$ can be regarded as the transition function of some vector bundle on the Riemann sphere with coordinate neighborhoods $CP^1 \setminus \{0\}$ and C. According to the Birkhoff–Grothendieck theorem (see (1.4.14)), there exist a matrix Γ holomorphically invertible in C and a matrix $U(z)$ holomorphically invertible in $CP^1 \setminus \{0\}$ such that

$$\Gamma(z)T(z) = z^K U(z), \tag{6.4}$$

where $K = \operatorname{diag}(k_1, \ldots, k_p)$, $k_1 \geq \cdots \geq k_p$.

PROPOSITION 4.6.1. *The tuple of numbers $-k_1, \ldots, -k_p$ coincides with the splitting type of the canonical bundle F^0 constructed from the monodromy representation of the original system* (6.1).

PROOF. Since the normalizations of the system constructed are equal to zero at all the points $a_1, \ldots, a_n$, a fundamental matrix $Z(y)$ of this system can be regarded as the transition function of the bundle F^0 with coordinate neighborhoods O and C, where O is a neighborhood of ∞. But by construction $Z(y) = T(z)Y(y)$, where $Y(y)$ is the fundamental matrix of the original system. (More precisely, it is a branch of that matrix; we recall here that ∞ is a point of holomorphy of the original system.)

By (6.4),

$$\Gamma(z)Z(y)Y^{-1}U^{-1} = z^K$$

at O. Consequently, the diagonal elements of the matrix $-K$ really do determine the splitting type of F^0.

REMARK 4.6.1. Since $T(y)$ is a known meromorphic matrix, K can be algorithmically computed by the procedure described in Lemma 11.2 in [26].

PROPOSITION 4.6.2. *Consider a Fuchsian system* (6.1) *with normalizations φ_i^j equal to zero for $i = 2, \ldots, n$ and $j = 1, \ldots, p$, and suppose that the matrix B_1 is reduced to Jordan normal form. Assume that the eigenvalues of B_1 (that is, the elements β_1^j on the diagonal of B_1) satisfy the inequalities $\varphi_1^1 \geq \cdots \geq \varphi_1^p$ (recall that $\varphi_i^j = [\operatorname{Re} \beta_i^j]$).*

The tuple $\varphi_1^1, \dots, \varphi_1^p$ *coincides with the splitting type of the canonical extension* F^0 *constructed from the monodromy of the original system if and only if the transformation*

$$g(y) = (z - a_1)^{-\Lambda_1} f(y) \tag{6.5}$$

with the matrix $\Lambda_1 = \operatorname{diag}(\varphi_1^1, \dots, \varphi_1^p)$ *carries the original system into a system that, as before, is Fuchsian at* a_1.

COROLLARY 4.6.1. *Suppose that the system* (6.1) *satisfies the conditions of Proposition* 4.6.2. *The tuple* $\varphi_1^1, \dots, \varphi_1^p$ *coincides with the splitting type of the canonical extension* F^0 *constructed from the monodromy of the original system if and only if*

$$\sum_{i=2}^{n} \frac{b_i^{kl}}{(a_i - a_1)^s} = 0, \quad \textit{for } 0 < s < \lambda_1^k - \lambda_1^l, \ k < l, \tag{6.6}$$

where $B_i = ((b_i^{kl}))$.

PROOF OF THE COROLLARY. It follows from (1.3.2) that the new system is Fuchsian at a_1 if and only if the matrices

$$\sum_{i=1}^{n} (z - a_1)^{-\Lambda_1} \frac{B_i}{z - a_i} (z - a_1)^{\Lambda_1}$$

have simple poles at a_1. Expanding $(z - a_i)^{-1}$ in a power series with respect to $(z - a_i)$, we get that the last condition is equivalent to (6.6).

PROOF OF PROPOSITION 4.6.2. *Sufficiency.* It follows from the condition of the proposition and the form of the transformation (6.5) that all the normalizations of the new system at all the points $a_1, \dots, a_n$ are equal to zero and that a branch of its fundamental matrix has at ∞ the form $T'(z) = z^{-\Lambda_1} T(z)$, where $T(z)$ is holomorphically invertible in a neighborhood of ∞. The sufficiency now follows from Proposition 4.6.1.

Necessity. We consider the decomposition (1.2.6) for our system. It follows from the form of the matrix B_1 of our system, from the same form of the matrix L_1 in (1.2.3) (recall that the notation Λ_i is being used in this chapter instead of A_i), and from (1.2.20) that

$$U_1(a_1) = \begin{pmatrix} P_1 & & 0 \\ & \ddots & \\ 0 & & P_s \end{pmatrix}, \tag{6.7}$$

where each $k_j \times k_j$ block P_j relates to a block of B_1 corresponding to a single eigenvalue, and the eigenvalues corresponding to distinct P_j and P_m are distinct. To prove the necessity it suffices to prove the following assertion.

LEMMA 4.6.1. *The matrix* $U_1(z)$ *satisfies the equality*

$$U_1(z)(z - a_1)^{\Lambda_1} = (z - a_1)^{\Lambda_1} U_1^0(z), \tag{6.8}$$

where $U_1^0(z)$ *is holomorphically invertible at* a_1.

PROOF. It can be assumed without loss of generality that $a_1 = 0$. Just as in the paragraph preceding Proposition 4.6.1, we conclude that $U_1(z)z^{\Lambda_1}$ can be regarded as the transition function of the bundle F^0 with coordinate neighborhoods O (a neighborhood of zero) and $CP^1 \setminus \{0\}$. Therefore, there exists a matrix $\Gamma(z)$ holomorphically invertible away from zero and a matrix $U'(z)$ holomorphically invertible in O such that

$$\Gamma(z)U_1(z)z^{\Lambda_1} = z^{\Lambda_1}U'(z). \tag{6.9}$$

We show that the principal minors of $U'(0)$ of orders $k_1, \dots, k_s$ are all nonzero. Consider the first principal minor (of order k_1) and assume that it is equal to zero. In this case there is a nonsingular matrix

$$R = \begin{pmatrix} R_1 & 0 \\ 0 & I^{p-k_1} \end{pmatrix},$$

(where I^{p-k_1} is the $(p-k_1)\times(p-k_1)$ identity matrix) such that

$$R\Gamma(z)U_1(z)z^{\Lambda_1} = z^{\Lambda_1}U''(z), \tag{6.10}$$

and the first k_1 elements of the first row of $U''(z)$ vanish at zero. But in this case the first row of the matrix $x^{\Lambda_1}U''(x)x^{-\Lambda_1}$ also vanishes at zero, because $\varphi_1^k > \varphi_1^l$ for $k \le k_1$ and $l > k_1$. Thus, it follows from (6.10) that the first row of the matrix

$$R\Gamma(z) = z^{\Lambda_1}U''(z)z^{-\Lambda_1}U_1(z)$$

vanishes at zero. Since $R\Gamma(z)$ is holomorphically invertible away from zero, we get that this first row is identically equal to zero, and hence $\det\Gamma(x) \equiv 0$, which is impossible. But this, in turn, means that the principal minor of order k_1 is nonzero.

Suppose that the principal minors of orders $k_1, \dots, k_m$ are all nonzero. We show that the same is true for the principal minor of order k_{m+1}. Assume the contrary. In this case there is a nonsingular upper triangular matrix R' of the form

$$R' = \begin{pmatrix} R_1' & 0 \\ 0 & I^{p-t} \end{pmatrix}$$

(where $t = \sum_{i=1}^{m+1} k_i$) such that $R'\Gamma(z)U_1(z)z^{\Lambda_1} = z^{\Lambda_1}U''(z)$, and the first t elements of some row of $U''(x)$ with index j $(t-k_m < j \le t)$ vanishes at zero. Just as above, it is not hard to show that in this case the jth row of the matrix $R'\Gamma(z)$ vanishes at zero, which is impossible. Consequently, this minor is also nonzero.

Since the principal minors of orders $k_1, \dots, k_s$ are all nonzero, there is a nonsingular constant matrix R'' of the form

$$R'' = \begin{pmatrix} R_1'' & & * \\ & \ddots & \\ 0 & & R_s'' \end{pmatrix}$$

such that $R''\Gamma(z)U_1(z)z^{\Lambda_1} = z^{\Lambda_1}\widetilde{U}(z)$, and all the principal minors of $\widetilde{U}(x)$ are nonvanishing at zero. Therefore, by Lemma 4.1.2 and Corollary 4.1.1, there exists a matrix Γ_1 that is holomorphically invertible away from zero and such that

$$\Gamma_1(z)R''\Gamma(z)U_1(z)z^{\Lambda_1} = z^{\Lambda_1}V^0(z) = V(z)z^{\Lambda_1}, \tag{6.11}$$

where $V(0)$ is an upper triangular matrix, and $V(z)$ and $V^0(z)$ are holomorphically invertible at zero. Denote the matrix $\Gamma_1R''\Gamma$ by Γ'. By construction, this matrix is holomorphically invertible on the whole Riemann sphere with the exception of the point zero. But it follows from (6.11) that $\Gamma'(z) = V(z)U_1^{-1}(z)$ is also holomorphically invertible at zero. Therefore, $\Gamma_1(z) \equiv \text{const}$. Also,

$$\Gamma' = V(0)U_1^{-1}(0) = \left[\lim_{z\to 0} z^{\Lambda_1} V^0(0) z^{-\Lambda_1}\right] U_1^{-1}(0) \tag{6.12}$$

has the form

$$\Gamma' = \begin{pmatrix} Q_1 & & 0 \\ & \ddots & \\ 0 & & Q_s \end{pmatrix},$$

where the order of the matrix Q_j is k_j. Indeed, the block upper triangular form of Γ' follows from the first equality in (6.12) (since $V(0)$ and $U_1^{-1}(0)$ have analogous form), and the block lower triangular form of Γ' follows from the second equality in (6.12), because $U_1^{-1}(0)$ has similar form, and $\varphi_1^j > \varphi_1^m$ for $j \le k_1 + \dots + k_r$ and $m > k_1 + \dots + k_r$, $r = 1, \dots, s-1$.

It follows from the form of Γ' that this matrix commutes with the matrix z^Λ. Therefore, it follows from (6.11) that $U_1(z)z^{\Lambda_1} = z^{\Lambda_1}(\Gamma')^{-1}V^0(z)$. Denote the matrix $(\Gamma')^{-1}V^0(x)$ by $U_1^0(x)$. The lemma and the proposition are proved.

COROLLARY 4.6.2. *Suppose that all the normalizations of a Fuchsian system* (6.1) *at the points* $a_2, \dots, a_n$ *are zero, and the matrix* B_1 *has a Jordan normal form such that the exponents* β_1^j *on the diagonal of* B_1 *are arranged in the nonincreasing order of the normalizations* $\varphi_1^j = [\operatorname{Re}\beta_1^j]$. *Then the tuple* $\varphi_1^1, \dots, \varphi_1^p$ *coincides with the splitting type of the canonical extension* F^0 *if and only if the conditions*

$$u_1^{lm}(z) = (z-a_1)^{\lambda_1^l - \lambda_1^m}(c^{lm} + o(1)), \qquad l > m,\ c^{lm} \in C, \tag{6.13}$$

hold for the elements u_1^{lm} *of the matrix* $U_1(z)$ *in the Levelt decomposition* (1.2.6).

The proof follows immediately from (6.8).

EXAMPLE 4.6.1. The Fuchsian system

$$\frac{df}{dz} = \left[\begin{pmatrix} 1 & 0 & 0 \\ 0 & 0 & 0 \\ 0 & 0 & -1 \end{pmatrix}\frac{1}{z} + \begin{pmatrix} -1 & -1/2 & 0 \\ 1 & 1 & -1/2 \\ 2 & 1 & 0 \end{pmatrix}\frac{1}{z+1} + \begin{pmatrix} 0 & 1/2 & 0 \\ -1 & -1 & 1/2 \\ -2 & -1 & 1 \end{pmatrix}\frac{1}{z-1}\right] f \tag{6.14}$$

has zero normalizations at the points -1 and 1. The splitting type of the canonical extension F^0 constructed from the monodromy of this system is equal to $1, 0, -1$. Each monodromy matrix $\chi(g_i)$ of this system can be reduced to a Jordan normal form consisting of a single block with eigenvalue 1 of multiplicity three. The monodromy representation of this system is irreducible.

PROOF. The matrices B_2 and B_3 of the system (6.14) have zero eigenvalues, and their rank is equal to two. This means that the normalizations at the points -1 and 1 are equal to zero, and each matrix B_i, $i = 2, 3$, can be reduced to a Jordan block with zero eigenvalue. It follows from (1.2.3), (1.2.5), and (1.2.11) that for these matrices the matrices E_i are similar to B_i, $i = 2, 3$, and therefore the matrices $\chi(g_i)$ can be reduced to a Jordan block with the eigenvalue 1.

The substitution

$$g(y) = z^{\begin{pmatrix} -1 & 0 & 0 \\ 0 & 0 & 0 \\ 0 & 0 & 1 \end{pmatrix}} f(y)$$

carries the original system to the system

$$\frac{dg}{dz} = \left[\begin{pmatrix} 0 & -1 & 0 \\ 0 & 0 & -1 \\ 0 & 0 & 0 \end{pmatrix} \frac{1}{z} + \text{the holomorphic part at } 0 \right] g.$$

It follows from Proposition 4.6.2 that the triple of numbers $1, 0, -1$ is the splitting type of F^0. By starting from the form of the coefficient matrix it can be proved exactly as above that the monodromy matrix $\chi(g_1)$ can be reduced to a Jordan block with eigenvalue 1.

It follows from results in Chapter I that the monodromy of the system (6.14) is irreducible, because a reducible monodromy with the properties we have established cannot be the monodromy of a Fuchsian system (see Theorem 1.4.2).

REMARK 4.6.2. It follows from Lemma 4.1.3, (4.1.11), and Proposition 4.6.1 that any Fuchsian system whose normalizations are zero at all but one singular point has the following property. The maximal normalization of this system at this last singular point cannot be less than k_1, and the minimal normalization does not exceed k_p, where $k_1, \dots, k_p$ constitute the splitting type of the canonical bundle F^0 constructed from the monodromy of the system. The following two assertions are theorems on realization of the splitting type.

THEOREM 4.6.1. *For any holomorphic vector bundle E on the Riemann sphere with splitting type $k_1, \dots, k_p$ and for any collection of points $a_1, \dots, a_n$ with $n = \max(\max_{i=1,\dots,p-1}((k_i - k_{i+1}) + 2), 3)$ there exists a Fuchsian system with singular points $a_1, \dots, a_n$ and monodromy χ such that the following conditions hold*:

i) *χ is irreducible.*

ii) *The canonical extension F^0 of the vector bundle F constructed from the monodromy representation χ has splitting type $k_1 - s, \dots, k_p - s$, where $s - 1$ is equal to the integer part of the number $(1/p)(k_1 + \dots + k_p)$ if this number is not an integer, and $s = (1/p)(k_1 + \dots + k_p)$ if it is an integer.*

iii) *Each monodromy matrix $\chi(g_i)$ can be reduced to Jordan normal form consisting of a single Jordan block.*

There does not exist a Fuchsian system with the properties i) *and* ii) *with fewer than n singular points.*

COROLLARY 4.6.3. *For any holomorphic vector bundle E on the Riemann sphere with splitting type $k_1, \dots, k_p$ there exists a meromorphic connection with logarithmic singularities and irreducible monodromy whose number of singular points is equal to $n = \max(\max_{i=1,\dots,p-1}((k_i - k_{i+1}) + 2, 3)$. Moreover, there does not exist a connection with these properties and with fewer than n singular points.*

PROOF. Consider first the case $k_1 + \dots + k_p = 0$. Let $k_1, \dots, k_p$ be the splitting type of the bundle E and let $n = \max(\max_{i=1,\dots,p-1}((k_i - k_{i+1}) + 2, 3)$. We consider n arbitrary points $a_1, \dots, a_n$ on the Riemann sphere along with matrices

$B_1, \dots, B_n$ with elements b_i^{kl} satisfying the following conditions:

$$b_i^{kl} = 0, \qquad i = 2, \dots, n, \ k < l+1;$$

$$\sum_{i=1}^{n} b_i^{kl} = 0, \qquad 1 \le k, l \le p;$$

(6.15)
$$\sum_{i=2}^{n} \frac{b_i^{ll+1}}{a_i - a_1)^s} = 0, \qquad s = 1, \dots, n-1, \ s \ne k_l - k_{l+1};$$

$$b_1^{kl} = 0, \quad k \ne l, \qquad b_1^{ll} = k_l, \quad l = 1, \dots, p;$$

$$\sum_{i=2}^{n} \frac{b^{ll+1}}{(a_i - a_1)^{k_l - k_{l+1}}} \ne 0.$$

To these conditions we adjoin $(n-1)p$ more equations (the conditions that all the eigenvalues of the matrices $B_2, \dots, B_n$ be equal to zero). As a result we get

$$(n-1)\frac{(p-1)(p-2)}{2} + p^2 + p^2 + (n-1)p = \frac{n+3}{2}p^2 - \frac{(n-1)(p-2)}{2}$$

equations for np^2 unknowns. The difference d between these two numbers s

(6.16)
$$d = np^2 - \frac{n+3}{2}p^2 - \frac{(n-1)(p-2)}{2} = \frac{n-3}{2}p^2 + \frac{(n-1)(p-2)}{2}.$$

Thus, for $n \ge 3$ and $p \ge 2$ this difference is nonnegative (and grows rapidly with increasing n and p). Therefore, for these n and p there always exist matrices $B_1, \dots, B_n$ satisfying the system (6.15) and such that $B_2, \dots, B_n$ have all eigenvalues equal to zero (we omit here the technical proof of the consistency of all these equations). We consider the Fuchsian system (6.1) with the matrices obtained.

It follows from the first and third equations and the last inequality in (6.15) that $b_i^{ll+1} \ne 0$, $i = 2, \dots, n$. Therefore, $\operatorname{rank} B_i = p-1$, $i = 2, \dots, n$, since the determinant of the matrix obtained by omitting the first row and the last column of B_i, $i \ne 1$, is equal to $b_i^{12} \cdot \dots \cdot b_i^{p-1\,p} \ne 0$. By proceeding as in Example 4.6.1 it is not hard to show that each matrix $\chi(g_i)$, $i = 2, \dots, n$, can be reduced to Jordan normal form consisting of a single Jordan block, and that the eigenvalues of these matrices are equal to 1.

It follows from (6.15) that we are under the conditions of Corollary 4.6.1, and therefore the numbers $k_1, \dots, k_p$ determine the splitting type of F^0. Proceeding as in Example 4.6.1, we get in view of the last inequality in (6.15) that the transformation $g(y) = (z - a_1)^K f(y)$ with $K = \operatorname{diag}(-k_1, \dots, -k_p)$ carries our system into a system that is Fuchsian at the point a_1, whose coefficient matrix is

$$B_1' = \begin{pmatrix} 0 & t_1 & \cdots & 0 \\ \vdots & \ddots & \ddots & \vdots \\ 0 & \cdots & 0 & t_{p-1} \\ 0 & \cdots & 0 & 0 \end{pmatrix}.$$

Therefore, the monodromy matrix $\chi(g_1) = \exp(2\pi i B_1')$ reduces to a Jordan block with eigenvalue 1.

Just as in Example 4.6.1, we get that the monodromy of the original system is irreducible (for otherwise it could not be the monodromy of a Fuchsian system).

Let us consider the case $k_1 + \cdots + k_p \neq 0$. Suppose that $(1/p)(k_1 + \cdots + k_p)$ is not an integer, and let $\rho = (1/p)(k_1 + \cdots + k_p) - s + 1$. Then $0 < \rho < 1$, and $p(\rho + s - 1) = k_1 + \cdots + k_p$. We replace the matrix B_2 constructed above by a matrix B_2' satisfying the same conditions as B_2 with the exception of the following p conditions: B_2' has p eigenvalues equal to ρ. We also replace the fourth equation in (6.15) by

$$b_1^{kl} = 0, \quad k \neq l, \qquad b_1^{ll} = k_l - s, \quad l = 1, \dots, p. \tag{6.17}$$

After this, the second and fourth equations in (6.15) will again be consistent. The rest of the proof repeats the proof of the case $k_1 + \cdots + k_p = 0$.

If the number $(1/p)(k_1 + \cdots + k_p)$ is an integer, then it is only necessary to replace the fourth equation in (6.15) by (6.17).

We prove the second assertion of the theorem. Assume that there exists a Fuchsian system with $k < n$ singular points $a_2, \dots, a_{k+1}$ satisfying the conditions i) and ii) and such that the splitting type $k_1, \dots, k_p$ of the canonical extension constructed from its monodromy satisfies for some l the inequality

$$k < k_l - l_{l+1} + 2. \tag{6.18}$$

Proceeding as in the beginning of the section, we reduce this system to a system satisfying the conditions of Proposition 4.6.1 at the points $a_1 = \infty, a_2, \dots, a_k$. (The new system will not be Fuchsian at ∞, but its fundamental matrix at this point will have the form (6.4), with $-K = \operatorname{diag}(k_1, \dots, k_p)$.) Therefore, the substitution

$$g(y) = z^K f(y) \tag{6.19}$$

carries our system into a system for which ∞ is a point of holomorphy. The latter means that the matrix

$$z^K \left(\sum_{i=2}^{k+1} \frac{B_i}{z - a_i} \right) z^{-K} + \frac{K}{z} \tag{6.20}$$

is holomorphic at ∞. Proceeding just as in the proof of Corollary 4.6.1 (with a_1 replaced by ∞), we get that the elements b_i^{kl} of the matrices B_i satisfy the equations

$$\sum_{i=2}^{k+1} a_i^s b_i^{mt} = 0, \qquad s = 0, \dots, k_m - k_t, \quad m < t. \tag{6.21}$$

Since $k < k_l - k_t + 2$ for all $t > l$ by assumption, we get that the elements b_i^{mt} are equal to zero for all $m \leq l$ and $t > l$ (since for each fixed $t > l$ they satisfy the system (6.21) of k linearly independent homogeneous equations). It follows from Lemma 1.3.5 that the monodromy of this system is reducible. We get a contradiction to condition i) of the theorem. The second assertion of the theorem is also proved.

EXAMPLE 4.6.2. Let $p = 2$, $n = 3$, $k_1 = 0$, and $k_2 = -1$ (the first nontrivial case; note that $d = 0$ in this case). The system

$$\frac{df}{dz} \left[\begin{pmatrix} 0 & 0 \\ 0 & -1 \end{pmatrix} \frac{1}{z} + \begin{pmatrix} -1/4 & 1/2 \\ -1/8 & 1/4 \end{pmatrix} \frac{1}{z+1} + \begin{pmatrix} 1/4 & -1/2 \\ 1/8 & 3/4 \end{pmatrix} \frac{1}{z-1} \right] f \tag{6.22}$$

satisfies all the conditions of the preceding theorem, and the canonical extension constructed from its monodromy has splitting type $0, -1$.

The case $p = 3$, $n = 3$, $k_1 = 1$, $k_2 = 0$, $k_3 = -1$ was presented in Example 4.6.1, and the case $p = 2$, $n = 4$, $k_1 = 1$, $k_2 = -1$ in Example 2.2.1.

REMARK 4.6.3. As a corollary to the preceding theorem we get a second proof of (4.2.1). Indeed, it follows from Corollary 4.6.3 that for each vector bundle E with a meromorphic connection ∇ having only logarithmic singularities the number n of singular points satisfies the inequality

$$k_1 - k_i \leq \sum_{j=1}^{i-1}(k_j - k_{j+1}) \leq (i-1) \max_{i=1,\dots,p-1}(k_i - k_{i+1}) \leq (i-1)(n-2).$$

Therefore,

$$\sum_{i=1}^{p}(k_1 - k_i) \leq (n-2)p(p-1)/2,$$

and the inequality (4.2.1) is satisfied.

CHAPTER V

Reducible Representations

All cases of a negative solution of the Riemann–Hilbert problem involve reducibility of the monodromy. Along with new results, this chapter contains generalizations of results obtained for reducible systems in Chapters I and II to the case of an arbitrary number of equations.

1. Reducibility and regularity

The following theorem, which is a generalization of Lemma 1.3.8 (see also [7]), establishes a connection between the reducibility of the monodromy of a system with regular singular points and the reducibility of its coefficient matrix.

THEOREM 5.1.1. *Suppose that the representation χ is reducible, and let χ_1 be a subrepresentation of χ of dimension l. There exists a system* (1.1.1) *with the given monodromy χ and with regular singular points $a_1, \dots, a_n$ that is Fuchsian at all but perhaps one of these points and has a fundamental matrix $T(y)$ of the form*

$$T(y) = \begin{pmatrix} T^1 & * \\ 0 & T^2 \end{pmatrix}, \tag{1.1}$$

where $T^1(y)$ is an $l \times l$ matrix-valued function that is invariant under the action of the monodromy χ_1.

PROOF. We consider a system (1.1.1) with monodromy χ that is Fuchsian at $a_2, \dots, a_n$ and regular at a_1. (The existence of such a system was proved in [52]; see also Proposition 1.4.1 in this monograph.)

Let $T(y)$ be a fundamental matrix of the solution space X of this system with the first l columns forming an associated basis of an l-dimensional subspace $X_l \subset X$ that is invariant under the action of the monodromy χ_1. Then

$$T(y) = (T_l(y)|*). \tag{1.2}$$

Proceeding as at the beginning of the proof of Theorem 4.3.1 (up to the formula (4.3.1)), we can assume that $T_l(y)$ has in $\widetilde{O}_1$ the form

$$T_l(y) = (z - a_1)^C V^l(z)(z - a_1)^{\Lambda_1^l}(y - a_1)^{E_1^l}, \tag{1.3}$$

where Λ_1^l is the matrix of normalizations of X_l, $E_1^l = (2\pi i)^{-1} \ln \chi_1(\sigma_1)$, and $\operatorname{rank} V^l(a_1) = l$.

Denote by Y_l a matrix with nonzero determinant formed by l rows of $T_l(y)$. The space X' generated by the rows of Y_l is the solution space of a system of the form

(1.1.1) with coefficient matrix $\omega_l = dY_lY_l^{-1}$. This system has monodromy χ_1 and singular points $a_1, \dots, a_n$, and perhaps also some additional false singular points $a_1', \dots, a_k'$. These are points for which $\det Y_l(\pi^{-1}(z)) = 0$ (see §3 in Chapter I and §4 in Chapter IV). Therefore, the sum s' of the exponents of X' at these points is nonnegative:

$$s' \geq 0. \tag{1.4}$$

For the sum s'' of the exponents of the space X' at $a_1, \dots, a_n$ we have the inequality

$$s'' \geq s_l, \tag{1.5}$$

where s_l is the sum of the exponents of the space X_l (see the analogous assertion (1.3.18) in Chapter I).

By the residue theorem, we get as in Proposition 1.2.3 that for the form $\operatorname{tr}\omega$

$$s_c + s'' + s' = 0, \tag{1.6}$$

where s_c is the sum of the diagonal elements of C from the rows with the same indices as the rows appearing in Y_l. From (1.4)–(1.6),

$$s_c \leq -s_l. \tag{1.7}$$

It can be assumed without loss of generality that Y_l is formed by the first rows of $T_l(y)$. (Otherwise we permute the rows of T_l, which corresponds to passing from T_l to a matrix ST_l, where $\det S \neq 0$.) Denote by V_l^l the matrix formed by the rows of V^l with indices $1, \dots, l$. The matrix $(z-a_1)^C V^l(z)$ in (1.3) has the form required in Lemma 4.1.1 (with $Y(z)$ replaced by V_l^l).

Therefore, there is a matrix $\Gamma(z)$ such that

$$\Gamma(z)(z-a_1)^C V^l(z) = (z-a_1)^{C'} \begin{pmatrix} V_l^l(z) \\ W'(z) \end{pmatrix}$$

with the corresponding conditions on C' (see Lemma 4.1.1).

Moreover, $\Gamma(z)$ can be chosen so that one of the following two conditions holds for any row $w_m'(z)$ of the matrix $W'(z)$:

a) $w_m'(z) \equiv 0$;

b) $w_m'(a_i) = -\sum_{r=1}^k d_{j_r} y_{j_r}(a_i)$, where $\prod_{r=1}^k d_{j_r} \neq 0$ and $\min(c_{j_1}, \dots, c_{j_k}) = c_{j_s} < c_m'$.

Indeed, if the inequality $c_{j_s} \geq c_m'$ holds for the row $w_m'(x)$, then we again use the procedure (4.1.2), and so on, until we get a) or b); then we construct the corresponding Q_j^m in (4.1.3).

Assume that b) holds for some row w_m' of the matrix W' (and, correspondingly, a) does not hold). In this case, by transposing the rows with indices j_s and $l+m$ in the matrix

$$(z-a_1)^{C'} \begin{pmatrix} V_l^l(z) \\ W'(z) \end{pmatrix},$$

we get a matrix

$$(z-a_1)^{C_1} \begin{pmatrix} \widetilde{V}_l^l(z) \\ W''(z) \end{pmatrix}$$

with $\det \widetilde{V}_l^l(a_1) \neq 0$.

Under such a transformation the matrix $T_l(y)$ passes into the matrix

$$T_l'(y) = \Gamma(z)T_l(y) = (z-a_1)^{C_1}\widetilde{V}^l(z)(z-a_1)^{\Lambda_1'}(y-a_1)^{E_1^l}. \tag{1.8}$$

It follows from the condition b) that $s_{c_1} > s_c$, where s_{c_1} is the sum of the first l diagonal elements of C_1. Moreover, since $\det\Gamma(z) \neq 0$ for $z \neq a_1$, we get that $s_l' \geq s_l$ for the space generated by the columns of T_l'. From (1.6) and the last inequalities,

$$s_c < s_{c_1} \leq -s_l' \leq -s_l. \tag{1.9}$$

If b) holds for some row of $W''(x)$, then we again use the procedure described above, and so on. It follows from (1.9) that in finitely many steps we get a matrix $\Gamma_0(z)(z-a_1)^C V^l(z)$ whose columns all satisfy the condition a).

We pass from the matrix $T(y)$ in (1.2) to $\widetilde{T}(y) = \Gamma_0(z)T(y)$ and get a matrix of the form (1.1). Since $\det\Gamma_0(z) \neq 0$ for $z \neq a_1$, it follows that $\widetilde{T}(y)$ is as before a fundamental matrix of a system that is Fuchsian at $a_2, , \dots, a_n$. The point a_1 is a regular singular point of this system, and it does not have other singular points.

COROLLARY 5.1.1. *Suppose that the monodromy matrices $G_i = \chi(\sigma_i)$ can be simultaneously reduced to the form*

$$G_i = \begin{pmatrix} \boxed{G_i^1} & & & * \\ 0 & \boxed{G_i^2} & & \\ \vdots & & \ddots & \\ 0 & 0 & \cdots & \boxed{G_i^k} \end{pmatrix}, \qquad i = 1, \dots, n, \tag{1.10}$$

where the matrices $G_1^j, \dots, G_n^j$ form an irreducible collection for all $j = 1, \dots, k$. There exists a regular system (1.1.1) *that is Fuchsian away from the point a_1 and has monodromy χ, and its coefficient matrix ω has the form* (1.10) (*with the G_i^j replaced by the corresponding ω^j*).

PROOF. Denote by χ_l the subrepresentation of χ defined by

$$\chi_l(\sigma_i) = \begin{pmatrix} \boxed{G_i^1} & & & * \\ 0 & \boxed{G_i^2} & & \\ \vdots & & \ddots & \\ 0 & 0 & \cdots & \boxed{G_i^l} \end{pmatrix}$$

According to Theorem 5.1.1, there exists a system with the given monodromy that has a fundamental matrix $T(z)$ of the form

$$T(z) = \begin{pmatrix} T^1(y) & * \\ 0 & \widetilde{T}(y) \end{pmatrix},$$

where $T^1(y)$ is a fundamental matrix of a system with monodromy χ_1. Therefore, $\widetilde{T}(y)$ is a fundamental matrix for a system (1.1.1) with monodromy χ/χ_1. We apply Theorem 5.1.1 to $\widetilde{T}(y)$, and so on. In k steps we get a matrix $T'(y)$ of the form (1.10). The coefficient matrix of a system with this fundamental matrix has the required form.

The analogous assertion for Fuchsian systems is not true in general.

PROPOSITION 5.1.1. *There are points $a_1, \dots, a_4$ on the Riemann sphere such that the monodromy representation χ in* (1.1.5) *with monodromy matrices G_i (in the order of their appearance; for the meaning of $\underline{1}$, see the proof)*

$$\begin{pmatrix} 1 & 1 & \underline{1} & -1 & 0 & 0 & -1 \\ 0 & -1 & -1 & 0 & 0 & 0 & \underline{1} \\ 0 & 0 & 1 & \underline{1} & 2 & 2 & 2 \\ 0 & 0 & 0 & 1 & \underline{1} & 0 & 1 \\ 0 & 0 & 0 & 0 & 1 & \underline{1} & 1 \\ 0 & 0 & 0 & 0 & 0 & 1 & 0 \\ 0 & 0 & 0 & 0 & 0 & 0 & -1 \end{pmatrix}, \quad \begin{pmatrix} 1 & 1 & 0 & \underline{1} & 1 & 1 & 0 \\ 0 & -1 & \underline{1} & 1 & -1 & 1 & -1 \\ 0 & 0 & -1 & -1 & \underline{1} & -1 & 0 \\ 0 & 0 & 0 & 1 & 1 & \underline{1} & 0 \\ 0 & 0 & 0 & 0 & -1 & -1 & \underline{1} \\ 0 & 0 & 0 & 0 & 0 & 1 & 0 \\ 0 & 0 & 0 & 0 & 0 & 0 & -1 \end{pmatrix},$$

$$\begin{pmatrix} 1 & 0 & \underline{1} & 0 & -1 & 0 & 0 \\ 0 & -1 & -1 & \underline{1} & -1 & 1 & 2 \\ 0 & 0 & 1 & 1 & -1 & \underline{1} & 2 \\ 0 & 0 & 0 & -1 & \underline{1} & -1 & -2 \\ 0 & 0 & 0 & 0 & -1 & 1 & \underline{1} \\ 0 & 0 & 0 & 0 & 0 & 1 & 1 \\ 0 & 0 & 0 & 0 & 0 & 0 & -1 \end{pmatrix}, \quad \begin{pmatrix} 1 & 0 & 1 & -1 & \underline{1} & 1 & 0 \\ 0 & -1 & \underline{1} & -2 & 0 & 0 & 0 \\ 0 & 0 & -1 & \underline{1} & 0 & 0 & 0 \\ 0 & 0 & 0 & -1 & 1 & 0 & \underline{1} \\ 0 & 0 & 0 & 0 & 1 & \underline{1} & 0 \\ 0 & 0 & 0 & 0 & 0 & 1 & 1 \\ 0 & 0 & 0 & 0 & 0 & 0 & -1 \end{pmatrix}$$

can be realized as the monodromy representation of some Fuchsian system, but cannot be realized as the monodromy representation of a Fuchsian system (1.1.1) *with upper triangular coefficient matrix ω.*

PROOF. We show first that there does not exist a Fuchsian system with the given monodromy whose coefficient matrices have upper triangular form. Assume the contrary: assume that such a system exists. Denote by $T(y)$ an upper triangular fundamental matrix of the solution space X of this system. Then the subspace X_l, $1 \le l < p$, generated by the first l columns $y_1, \dots, y_l$ of $T(y)$ is invariant under the action of the monodromy. Each matrix G_i can be reduced to a Jordan normal form consisting of two Jordan blocks with respective eigenvalues 1 and -1. Denote by e^i_j the vectors of the Jordan bases for the blocks with eigenvalues equal to 1, and by f^i_k those for the blocks with eigenvalues -1. Using only upper triangular transformations, we can reduce each G_i to a form obtained by a reshuffling of the two blocks. (This form can be obtained by zeroing out all the numbers above the diagonal in G_i except for those underlined.) The corresponding bases for G_i have the appearance

$$\begin{aligned} &e^1_1, f^1_1, e^1_2, e^1_3, e^1_4, e^1_5, f^1_2, && \text{for } G_1, \\ &e^2_1, f^2_1, f^2_2, e^2_2, f^2_3, e^2_3, f^2_4, && \text{for } G_2, \\ &e^3_1, f^3_1, e^3_2, f^3_2, f^3_3, e^3_3, f^3_4, && \text{for } G_3, \\ &e^4_1, f^4_1, f^4_2, f^4_3, e^4_2, e^4_3, f^4_4, && \text{for } G_4. \end{aligned}$$

We note that any of the vectors e^i_1, $i = 1, \dots, 4$, generates the subspace X_1, the vectors e^i_1 and f^i_1 generate the subspace X_2, and so on. Denote by a^i_j the normalizations $\varphi_i(e^i_j)$ of the vectors e^i_j, and by b^i_j the normalizations $\varphi_i(f^i_j)$. According to Example 1.1.3,

$$\begin{aligned} &a^1_1 \ge \cdots \ge a^1_5, \qquad b^1_1 \ge b^1_2, \\ &a^i_1 \ge \cdots \ge a^i_3, \qquad b^i_1 \ge \cdots \ge b^i_4, \qquad i = 2, 3, 4. \end{aligned} \tag{1.11}$$

Denote by s_l the sum of the exponents of X_l. Since each of the X_l is the solution space of a Fuchsian system of l equations by assumption, it follows from Proposition 1.2.3 that $s_l = 0$ for all $l = 1, \dots, 7$. By this equality and again by (1.11), $a_1^i = \cdots = a_5^i = a^i$ for some a_i, $i = 1, \dots, 4$, and $b_1^i = \cdots = b_4^i = b^i$, $i = 1, \dots, 4$.

Indeed, suppose, for example, that $b_1^i > b_4^i$ for some i. Then

$$b_1^1 + b_1^2 + b_1^3 + b_1^4 > b_2^1 + b_4^2 + b_4^3 + b_4^4.$$

Therefore,

$$0 = s_7 = s_6 + b_2^1 + b_4^2 + b_4^3 + b_4^4 + 2 < s_6 + b_1^1 + b_1^2 + b_1^3 + b_1^4 + 2 = s_6 + s_2 = 0,$$

which is impossible. (Recall that $s_2 = s_1 + b_1^1 + b_1^2 + b_1^3 + b_1^4 + 2$, and the term 2 appears because $\rho_i^2 = 1/2$ for the eigenvalues -1 of the monodromy.)

In this case the condition (1.2.22) of Proposition 1.2.3 has the form

$$\begin{aligned} 0 = s_7 &= 5a^1 + 3a^2 + 3a^3 + 3a^4 + 2b^1 + 1 + 4b^2 + 2 + 4b^3 \\ &= 2 + 4b^4 + 2 = 2a^1 - 2b^1 - 1, \end{aligned} \tag{1.12}$$

because $s_1 = s_2 = 0$. But it follows from (1.12) that $2a^1 - 2b^1 = 1$ for the integers a^1 and b^1, which is impossible. This contradiction shows that the given monodromy χ cannot be realized as the monodromy of a Fuchsian system with upper triangular coefficient matrix.

We now show that for some choice of the points $a_1, \dots, a_4$ the given representation can be realized by some Fuchsian system (of course, the coefficient matrix of such a system cannot be reduced to upper triangular form with preservation of the Fuchsian property of the system).

Let $a_1 = 0$, and let a_2, a_3, and a_4 be arbitrary points. According to Corollary 5.1.1, there exists a system (1.1.1) with the given monodromy χ that is Fuchsian at a_2, a_3, and a_4 and has a_1 as a regular singular point, and there exists a fundamental matrix $T(y) = ((t_{ij}))$ of this system with upper triangular form. By the residue theorem, applied to the elements t_{jj}, and by the fact that the eigenvalues of the monodromy operators for these elements are equal either to 1 or to -1, we get

$$\sum_{i=1}^{4} \varphi_i(t_{jj}) = 0, \quad j = 1, 6, \qquad \sum_{i=1}^{4} \varphi_i(t_{jj}) = -2, \quad j = 2, 7,$$

$$\sum_{i=1}^{4} \varphi_i(t_{jj}) = -1, \quad j = 3, 4, 5.$$

This system can be reduced by the $B(a_l, a_1, *, \dots, *)$ procedure (see (1.3.9)) to a system with fundamental matrix $T(y)$ (we keep the previous notation for the fundamental matrix of the new system) whose elements satisfy the conditions

$$\begin{aligned} &\varphi_i(t_{jj}) = 0, \quad j = 1, \dots, 7, \ i = 2, 3, 4, \qquad \varphi_i(t_{11}) = \varphi_i(t_{66}) = 0, \\ &\varphi_i(t_{22}) = \varphi_i(t_{77}) = -2, \qquad \varphi_i(t_{33}) = \varphi_i(t_{44}) = \varphi_i(t_{55}) = -1. \end{aligned} \tag{1.13}$$

We consider the two-dimensional subrepresentation χ_2 of χ and the corresponding subsystem of our system with fundamental matrix T_2. The decomposition (1.2.6) for this matrix in $\widetilde{O}_1$ has the form

$$T_2(y) = \begin{pmatrix} a(z) & r(z) \\ 0 & b(z) \end{pmatrix} z^{\begin{pmatrix} 0 & 0 \\ 0 & -2 \end{pmatrix}} y^{E_1^2}. \tag{1.14}$$

By using the $A(0, 2, 1, 0)$ procedure (see Lemma 1.3.2), we can ensure that the element $r(z)$ has the form $r(z) = cz^m(1 + o(1))$ in a neighborhood of zero, with $m > 0$. We remark that $c \neq 0$ in the last equality, for otherwise T_2 would be diagonal, and hence χ_2 would be commutative, which is not true. Therefore, just as in Proposition 2.5.2 it can be shown that for any point a_1' sufficiently close to zero the Fuchsian weight of χ_2 is equal to zero. (Formally, it is required in Proposition 2.5.2 that all the monodromy matrices be nondiagonalizable, but under the condition $c \neq 0$ the proof works also for our case without any changes.) We repeat all the preceding arguments with zero replaced by a_1' (which we again denote by a_1). Then the condition $m = 1$ holds in (1.14) (with z replaced by $z - a_1$), for otherwise the Fuchsian weight would be greater than zero by Proposition 2.2.2. We pass from our system to a system with fundamental matrix $T'(y) = \Gamma_1(z)T(y)$, where

$$\Gamma_1(z) = \begin{pmatrix} 1 & 0 & 0 \\ -\frac{b(a_1)}{c(z-a_1)} & 1 & 0 \\ \mathbf{0} & & I^5 \end{pmatrix}.$$

The matrix T_2' formed by intersecting the first two rows and columns of T' has the form

$$T_2'(y) = V_1^2(z)(z - a_1)^{-I}(y - a_1)^{E_1^2} \tag{1.15}$$

with a matrix

$$V_1^2(z) = \begin{pmatrix} a(z)(z - a_1) & c(1 + o(1)) \\ -\frac{a(z)b(a_1)}{c} & d(z) \end{pmatrix}$$

that is holomorphically invertible at a_1.

For the factor representation χ/χ_5 and the corresponding factor system we now do everything that we did above for χ_2. As a result we get a system (1.1.1) that is Fuchsian at a_2, a_3, and a_4 whose fundamental matrix on $\widetilde{O}_1$ will have a decomposition of the form (1.2.6) with matrix $A_1 = -I$ and

$$U_1(z) = \begin{pmatrix} V_1(z) & & * & & * \\ & u_{33} & * & * & \\ \mathbf{0} & 0 & u_{44} & * & * \\ & 0 & 0 & u_{55} & \\ 0 & & \mathbf{0} & & V_2(z) \end{pmatrix}, \tag{1.16}$$

and it will follow from (1.13) and (1.15) that the matrices $V_1(z)$ and $V_2(z)$ and the elements u_{33}, u_{44}, and u_{55} are holomorphically invertible at a_1. The elements denoted by $*$ in (1.16) will only be meromorphic at this point in general. Applying the $A(a_1, j, j, 0)$ procedure for $j = 3, 4, 5$ and the procedure in Lemma 4.1.1 to the matrix formed by the last two columns of $T'(y)$, we can ensure that the elements denoted by $*$ are holomorphic at a_1. The Fuchsian property of the system constructed at a_1 now follows from Corollary 1.2.2.

In some cases we can ensure the existence of a Fuchsian system with the same character of reducibility as the monodromy. The next theorem is a corollary to Lemma 4.1.1.

THEOREM 5.1.2. *Suppose the Fuchsian system* (1.1.1) *has reducible monodromy* χ, *and let* X_l *be a subspace of the solution space* X *of the system that is invariant under the action of the monodromy operators. If the sum* s_l *of the exponents of* X_l *over all singular points* $a_1, \dots, a_n$ *is equal to zero, then there exists a constant nonsingular matrix* S *such that the matrix* B' *of the Fuchsian system*

$$dg/dz = B'(z)g \tag{1.17}$$

obtained from (1.1.1) *by the substitution* $g = Sf$ *has the form* $b_{ij} = 0$ *for* $i = l+1, \dots, p$ *and* $j = 1, \dots, l$.

PROOF. We consider a fundamental matrix $T(y)$ of (1.1.1) with its first l columns forming a basis in X_l. Then $T(y) = (T_l(y)|*)$, according to (1.2).

For the corresponding matrix Y_l constructed in the proof of Theorem 5.1.1 and for the space X' we get from (1.4)–(1.6) and from the condition $s_l = 0$ that

$$s_l = s'' = s' = 0. \tag{1.18}$$

(Since the system (1.1.1) is Fuchsian, we have $C = 0$ in (1.3).) The matrix $V^l(x)$ has the form

$$V^l(z) = \begin{pmatrix} V_l^l(z) \\ W(z) \end{pmatrix}$$

in our case, and $\det V_l^l(a_1) \neq 0$, because the corresponding system with fundamental matrix $Y_l^* = V_l^l(z)(z-a_1)^{\Lambda_1^l}(y-a_1)^{E_1^l}$, $y \in \widetilde{O}_1$, is Fuchsian (this follows from the condition $s' = 0$ and Proposition 1.2.3). Since the rows of $V_l^l(a_1)$ form a basis in C^l, there exists a nonsingular constant matrix S such that

$$S\begin{pmatrix} V_l^l(a_1) \\ W(a_1) \end{pmatrix} = \begin{pmatrix} V_l^l(a_1) \\ 0 \end{pmatrix},$$

but the latter means that

$$SV^l(z) = (z-a_1)^C \begin{pmatrix} V_l^l(z) \\ W'(z) \end{pmatrix}, \tag{1.19}$$

where $C = \operatorname{diag}(0, \dots, 0, c_{l+1}, \dots, c_p)$, and $c_j > 0$ for $j = l+1, \dots, p$. We show that $W'(z) \equiv 0$. Assume the contrary. If $W'(z)$ is not identically zero, then, just as in the proof of Theorem 5.1.1 (after (1.7)), by interchanging some rows of SV^l with indices $j_s < l+1$ and $l+m$ we get a matrix

$$(z-a_1)^{C_1} \begin{pmatrix} \widetilde{V}_l^l(z) \\ W''(z) \end{pmatrix}$$

with $\det \widetilde{V}_l^l(a_1) \neq 0$ such that $s_{c_1} > s_c = 0$. But $s_{c_1} \leq -s_l = 0$ by virtue of (1.9) and (1.8), and this contradicts the preceding inequality. The latter means that $W'(z) \equiv 0$, and so the fundamental matrix $ST(y) = T'(y)$ of (1.17) has the form

$$T'(y) = {}^{l}\!\begin{pmatrix} * & * \\ \overline{0}| & * \end{pmatrix}_{l}$$

and hence $B'(z)$ has the form required in the theorem.

The next result follows immediately from Corollary 1.4.1 and Theorem 5.1.1.

COROLLARY 5.1.2. *Suppose that the monodromy* χ *of a Fuchsian system* (1.1.1) *is reducible, and that each monodromy matrix* $\chi(\sigma_i)$ *reduces to a Jordan normal form consisting of a single Jordan block. Then there exists a nonsingular constant matrix* S *such that the system* (1.17) *obtained from* (1.1.1) *by the substitution* $g = Sf$ *has the form* $b_{ij} = 0$ *for* $i = l+1, \ldots, p$ *and* $j = 1, \ldots, l$.

2. A new series of counterexamples

PROPOSITION 5.2.1. *Suppose that the matrices* $G_1, \ldots, G_n$ *have a common invariant subspace of dimension* l, $0 < l < p$, *satisfy the relation* $G_1 \cdot \ldots \cdot G_n = I$, *and are each reducible to a Jordan block with corresponding eigenvalue* ρ_i. *If the number* $\rho = \sum_{j=1}^{n} \frac{1}{2\pi i} \ln \rho_i$ *is not an integer, then the Riemann–Hilbert problem for the representation* (1.1.5) *with monodromy matrices* $G_i = \chi(\sigma_i)$ *and with arbitrary points* $a_1, \ldots, a_n$ *has a negative solution.*

PROOF. We consider the decomposition (1.4.14) for the canonical extension F^0 constructed from the given monodromy. The first Chern number for this bundle is equal to $c_1(F^0) = c_1^0 + \cdots + c_p^0 = pc_1^0$, since $c_1^0 = \cdots = c_p^0$ by Theorem 1.4.2. According to the same theorem (see (1.4.19)), if the given representation is realized by a Fuchsian system, then all the normalizations for this system are equal at each point (and are equal to some number φ_i). Therefore, from (1.4.13) and the fact that $\sum_{i=1}^{n} E_i^0 = p\rho$ it follows that $c_1(F^0) = p\rho + p\sum_{i=1}^{n} \varphi_i$.

Consequently,

$$p\rho = pc_1^0 = p\sum_{i=0}^{n} \varphi_i,$$

which immediately gives us that ρ is an integer.

EXAMPLE 5.2.1. We consider the matrices

$$G_1 = \begin{pmatrix} 1 & 1 & 0 & 0 \\ 0 & 1 & 1 & 0 \\ 0 & 0 & 1 & 1 \\ 0 & 0 & 0 & 1 \end{pmatrix}, \qquad G_2 = \begin{pmatrix} 3 & 1 & 1 & -1 \\ -4 & -1 & 1 & 2 \\ 0 & 0 & 3 & 1 \\ 0 & 0 & -4 & -1 \end{pmatrix},$$

$$G_3 = \begin{pmatrix} -1 & 0 & 2 & -1 \\ 4 & -1 & 0 & 1 \\ 0 & 0 & -1 & 0 \\ 0 & 0 & 4 & -1 \end{pmatrix}$$

and an arbitrary triple a_1, a_2, a_3 of points. A representation χ with singular points a_1, a_2, and a_3 and with monodromy matrices $\chi(\sigma_i) = G_i$, $i = 1, 2, 3$, cannot be the monodromy representation of a Fuchsian system.

PROOF. Note that $G_1 \cdot G_2 \cdot G_3 = I$, G_2 can be reduced to G_1, and G_3 can be reduced to a Jordan block with eigenvalue -1. Indeed, for G_2

$$S_2^{-1} G_2 S_2 = \begin{pmatrix} 1 & 1 & 0 & 0 \\ 0 & 1 & 1 & 0 \\ 0 & 0 & 1 & 1 \\ 0 & 0 & 0 & 1 \end{pmatrix}, \qquad S_2 = \frac{1}{3} \begin{pmatrix} 3 & 0 & -1 & 1 \\ -6 & 3 & -1 & 1 \\ 0 & 0 & 3 & 0 \\ 0 & 0 & -6 & 3 \end{pmatrix},$$

and for G_3

$$S_3^{-1}G_3S_3 = \begin{pmatrix} -1 & 1 & 0 & 0 \\ 0 & -1 & 1 & 0 \\ 0 & 0 & -1 & 1 \\ 0 & 0 & 0 & -1 \end{pmatrix}, \qquad S_3 = \frac{1}{2}\begin{pmatrix} 0 & 8 & 0 & 1 \\ 32 & -8 & 0 & 0 \\ 0 & 0 & 0 & -2 \\ 0 & 0 & -8 & -4 \end{pmatrix}.$$

The representation χ has a two-dimensional subrepresentation χ_1 with monodromy matrices G_i^1 that can be obtained by intersecting the first two rows and columns of G_i.

By the construction of E_i, $\rho_1 = \rho_2 = 0$ and $\rho_3 = (2\pi i)^{-1}\ln(-1) = 1/2$; therefore, the number $\rho = \rho_1 + \rho_2 + \rho_3 = 1/2$ is not an integer. According to Corollary 5.2.1, χ cannot be the monodromy representation of a Fuchsian system.

REMARK 5.2.1. All the counterexamples to the Riemann–Hilbert problem for representations of dimension three are unstable with respect to displacements of the singular points (see Proposition 2.5.3). In this sense Example 5.2.1 is the first (with respect to the dimension of the representation and the number of singular points) "stable" counterexample.

The upper triangularity of a representation does not ensure the solvability of the Riemann–Hilbert problem, as follows from the next example.

EXAMPLE 5.2.2. We consider an arbitrary quadruple a_1, a_2, a_3, a_4 of points and a representation (1.1.5) with monodromy matrices $G_i = \chi(\sigma_i)$ (in the order of their appearance)

$$\begin{pmatrix} 1 & 0 & \underline{1} & 0 & 0 & 1 & 0 \\ 0 & -1 & -1 & 0 & 0 & \underline{1} & 0 \\ 0 & 0 & 1 & \underline{1} & 2 & 2 & 1 \\ 0 & 0 & 0 & 1 & \underline{1} & 1 & 0 \\ 0 & 0 & 0 & 0 & 1 & 1 & \underline{1} \\ 0 & 0 & 0 & 0 & 0 & -1 & 0 \\ 0 & 0 & 0 & 0 & 0 & 0 & 1 \end{pmatrix}, \quad \begin{pmatrix} 1 & 0 & 1 & \underline{1} & -1 & 1 & 0 \\ 0 & -1 & \underline{1} & 1 & -1 & -1 & 0 \\ 0 & 0 & -1 & -1 & \underline{1} & 0 & 0 \\ 0 & 0 & 0 & 1 & 1 & 0 & \underline{1} \\ 0 & 0 & 0 & 0 & -1 & \underline{1} & -1 \\ 0 & 0 & 0 & 0 & 0 & -1 & 0 \\ 0 & 0 & 0 & 0 & 0 & 0 & 1 \end{pmatrix},$$

$$\begin{pmatrix} 1 & 0 & \underline{1} & 0 & -1 & 1 & 1 \\ 0 & -1 & -1 & \underline{1} & 0 & 1 & 0 \\ 0 & 0 & 1 & 1 & -1 & 2 & \underline{1} \\ 0 & 0 & 0 & -1 & \underline{1} & -2 & -1 \\ 0 & 0 & 0 & 0 & -1 & \underline{1} & 1 \\ 0 & 0 & 0 & 0 & 0 & -1 & 0 \\ 0 & 0 & 0 & 0 & 0 & 0 & 1 \end{pmatrix}, \quad \begin{pmatrix} 1 & 0 & 1 & -1 & \underline{1} & 0 & 0 \\ 0 & -1 & \underline{1} & -2 & 1 & 0 & 0 \\ 0 & 0 & -1 & \underline{1} & 0 & 0 & 0 \\ 0 & 0 & 0 & -1 & 1 & \underline{1} & 0 \\ 0 & 0 & 0 & 0 & 1 & -1 & \underline{1} \\ 0 & 0 & 0 & 0 & 0 & -1 & 0 \\ 0 & 0 & 0 & 0 & 0 & 0 & 1 \end{pmatrix}.$$

This representation cannot be the monodromy representation of a Fuchsian system.

PROOF. Assume the contrary. Let $T(y)$ be a fundamental matrix of the solution space X for a Fuchsian system (1.1.1) with the given monodromy. The proof now repeats that of Example 5.1.1 with the following changes. We must replace the scheme of associations of vectors in this example by the following:

$$\begin{array}{ll} e_1^1, f_1^1, e_2^1, e_3^1, e_4^1, f_2^1, e_5^1, & \text{for } G_1, \\ e_1^2, f_1^2, f_2^2, e_2^2, f_3^2, f_4^2, e_3^2, & \text{for } G_2, \\ e_1^3, f_1^3, e_2^3, f_2^3, f_3^3, f_4^3, e_3^3, & \text{for } G_3, \\ e_1^4, f_1^4, f_2^4, f_3^4, e_2^4, f_4^4, e_3^4, & \text{for } G_4. \end{array}$$

Then we must use the inequality $s_1 < 0$ (Lemma 1.3.7), which by Proposition 1.2.3 and (1.11) implies that

$$0 = s_7 = s_6 + a_5^1 + a_3^2 + a_3^3 + a_3^4 \leq s_6 + s_1 \leq 0.$$

Therefore, $s_6 = 0$ and $s_1 = a_5^1 + a_3^2 + a_3^3 + a_3^4 = 0$. From this equality and again from (1.11) we get that $a_1^i = \cdots = a_5^i = a^i$ and $b_1^i = \cdots = b_4^i = b^i$ for $i = 1, 2, 3, 4$. And we again arrive at the equality (1.12), which is impossible, whatever the integers a^1 and b^1.

3. Sufficient conditions for the Riemann–Hilbert problem to have a positive solution

For what follows we need a technical lemma.

LEMMA 5.3.1. *Suppose that the matrix $G_i = \chi(\sigma_i)$ of an irreducible representation χ has the form*

$$G_i = \begin{pmatrix} G_i' & 0 \\ 0 & G_i'' \end{pmatrix}, \tag{3.1}$$

where G_i' is a $t \times t$ matrix, $0 < t < p$. For any integers d_1 and d_2 there is a Fuchsian system (1.1.1) with monodromy χ whose normalizations φ_j^k satisfy the following conditions:

(i) $\varphi_j^k = d_1$, $j \neq i$, $k = 1, \ldots, p$;
(ii) $\varphi_i^k < d_2$, $k = 1, \ldots, t$;
(iii) $\varphi_i^m > d_2$, $m = t+1, \ldots, p$.

PROOF. It can be assumed without loss of generality that $i = 1$. We modify the proof of Theorem 4.3.1 as follows. Consider a regular system (1.1.1) with the given monodromy χ that is Fuchsian away from the point a_1 and whose normalizations φ_j^k satisfy condition i) of the lemma for $j \neq 1$. (The existence of this system follows from Theorem 4.3.2.) Moreover, it can be assumed that the matrix $\chi(\sigma_1)$ has the form (3.1) in the basis of columns of $T_1(y)$. (Otherwise, we transform the matrix $T_1(z)$ to $T_1(z)S$, where $S^{-1}\chi(\sigma_1)S$ has the form (3.1).) It follows from the assumptions of the lemma in this case that the upper triangular matrix E_1^0 in (4.3.1) has block form: $E_1^0 = \operatorname{diag}(E_1', E_2'')$, where $E_1' = \frac{1}{2\pi i}\ln G_1'$. We replace the matrix A_{φ_1} in (4.3.1) by $A_{\varphi_1}' = \operatorname{diag}(b_1', \ldots, b_p')$, where the numbers $b_1', \ldots, b_p'$ satisfy the conditions

$$\begin{gathered} b_1' < d_2 - \frac{(p-1)(n-2)}{2}, \qquad b_p' > d_2 + np + \frac{(p-1)^2(n-2)}{2}, \\ \sum_{j=1}^{p} b_j' = -pd_1(n-1), \\ {}_j - b_{j+1} > \frac{p^2(p-1)(n-2)}{2}, \qquad j = 1, \ldots, t-1, t+1, \ldots, p-1. \end{gathered} \tag{3.2}$$

Consider a system with the fundamental matrix $T_1'(y)$ constructed in Theorem 4.3.1. Since $\Phi = D + A_{\varphi_1}'$ has the block form $\Phi = \operatorname{diag}(\Phi', \Phi'')$, where the t diagonal elements of Φ' and the $p-t$ elements of Φ'' are arranged in nonincreasing order (in view of the last inequality in (3.2) and the inequality (4.3.3) of Theorem 4.3.1), it follows from the form of the matrix E_1^0 that the matrix-valued function

$$(z-a_1)^{\Phi} E_1^0 (z-a_1)^{-\Phi}$$

is holomorphic at a_1. Therefore, by Corollary 1.2.2 this system is Fuchsian at a_1.

From the third condition in (3.2) together with the condition $\sum_{i=1}^n \sum_{j=1}^p \beta_i^j = 0$ for the system constructed (see Proposition 1.2.3) and the inequality $0 \le \operatorname{Re} \rho_j^k < 1$ we get

$$-np < \operatorname{tr} C^{\varphi} \le 0, \tag{3.3}$$

where C^{φ} is from (4.3.1). Indeed,

$$\begin{aligned} 0 = \sum_{i=1}^n \sum_{j=1}^p \beta_i^j &= \sum_{i=2}^n \sum_{j=1}^p \beta_i^j + \sum_{j=1}^p \beta_1^j \\ &= pd_1(n-1) + \sum_{i=2}^n \sum_{j=1}^p \rho_i^j + \operatorname{tr}(D + A'_{\varphi_1}) + \sum_{j=1}^p \rho_1^j \\ &= \sum_{i=1}^n \sum_{j=1}^p \rho_i^j + \operatorname{tr} C^{\varphi} + pd_1(n-1) + \operatorname{tr} A'_{\varphi_1}, \end{aligned}$$

because $\operatorname{tr} C^{\varphi} = \operatorname{tr} D$ by virtue of (4.3.4). The inequality (3.3) now follows from the inequality $0 \le \sum_{i=1}^n \sum_{j=1}^p \rho_i^j < pn$ (which, in turn, follows from $0 \le \operatorname{Re} \rho_i^j < 1$) and from the third equality in (3.2): $\operatorname{tr} B' = \sum_{j=1}^p b'_j = -pd_1(n-1)$.

By (4.2.1) and (3.3),

$$\begin{gathered} c_1^{\varphi} p - \operatorname{tr} C^{\varphi} \le \frac{(n-2)p(p-1)}{2}, \\ c_1^{\varphi} p \le \frac{(n-2)p(p-1)}{2} + \operatorname{tr} C^{\varphi} \le \frac{(n-2)p(p-1)}{2}, \\ c_1^{\varphi} \le \frac{(n-2)(p-1)}{2}. \end{gathered} \tag{3.4}$$

It follows from (3.4) and the condition that $c_1^{\varphi}, \dots, c_p^{\varphi}$ are nonincreasing that

$$c_1^{\varphi} + \dots + c_{p-1}^{\varphi} \le c_1^{\varphi}(p-1) \le \frac{(n-2)(p-1)^2}{2}.$$

Therefore, from the left-hand inequality in (3.3),

$$c_p^{\varphi} > -np - \frac{(n-2)(p-1)^2}{2}. \tag{3.5}$$

According to (3.4), (3.5), and the first condition in (3.2),

$$b'_j + c_l^{\varphi} < b'_1 + c_1^{\varphi} < d_2$$

for all $j \le t$ and for all l; therefore, the first t diagonal elements in Φ do not exceed d_2. In precisely the same way the second condition in (3.2) gives us that

$$b'_j + c_l^{\varphi} > b'_p + c_p^{\varphi} > d_2$$

for all $j > t$ and for all l; therefore, the last $p-t$ diagonal elements of Φ are less than d_2. Thus, conditions ii) and iii) are also satisfied for the fundamental matrix $T'(y)$.

THEOREM 5.3.1. *Suppose that all the monodromy matrices $\chi(\sigma_j)$ of a representation χ reduce simultaneously to the form*

$$G_j = \begin{pmatrix} G_j^1 & * \\ 0 & G_j^2 \end{pmatrix}, \tag{3.6}$$

where each G_j^1 is an $l \times l$ matrix. Suppose that the collection of matrices $G_1^i, \dots, G_n^i$ determines a representation χ_i, $i = 1, 2$, and suppose that the representation χ_2 is realized as the monodromy of some Fuchsian system. If the representation χ_1 is irreducible and if the matrix G_i has the form (3.1) *with $t \le l$ for some i, then the monodromy χ also can be realized as the monodromy of some Fuchsian system.*

PROOF. It can be assumed without loss of generality that $i = 1$. We consider a fundamental matrix $Y_2(y)$ of a Fuchsian system with monodromy χ_2 such that $\chi_2(\sigma_j) = G_j^2$ in the basis of columns of $Y_2(y)$. Let

$$\max_{\substack{i=2,\dots,n;\\ j=1,\dots,p-l}} \varphi_i^j = d_1, \qquad \max_{j=1,\dots,p-l} \varphi_1^j = d_2 \tag{3.7}$$

for the normalizations of the system. We consider the corresponding fundamental matrix $Y_1(y)$ of the Fuchsian system constructed in the preceding lemma, with the monodromy χ_1 and with d_1 and d_2 in (3.7). Let $\chi_1(\sigma_j) = G_j^1$ in the basis of columns of $Y_1(y)$.

Denote by $T(y)$ a fundamental matrix of a regular system with monodromy χ that is Fuchsian at the singular points $a_2, \dots, a_n$. Suppose in addition that $T(y)$ has the form

$$T(y) = \begin{pmatrix} T^1 & * \\ 0 & T^2 \end{pmatrix}, \tag{3.8}$$

where T^1 and T^2 have monodromy G_j^1 and G_j^2, $j = 1, \dots, n$, respectively (see Theorem 5.1.1).

We consider the matrices $\Gamma_i = Y_i(y)(T^i(y))^{-1}$ and the new fundamental matrix $T'(y) = \Gamma(z)T(y)$ with

$$\Gamma = \begin{pmatrix} \Gamma_1 & 0 \\ 0 & \Gamma_2 \end{pmatrix}.$$

It follows from (3.7) that for the corresponding Levelt matrix $T_i'(y) = T'(y)S_i$, $i \ne 1$, in the decomposition (1.2.6) with a holomorphic matrix $V(z)$ and with $A_i = \operatorname{diag}(d_1, \dots, d_1, \varphi_i^1, \dots, \varphi_i^{p-1})$ the diagonal elements of A_i form a nonincreasing sequence. Hence, we get by Corollary 1.2.2 that the system (1.1.1) with fundamental matrix $T'(y)$ is Fuchsian at a_i for $i > 1$.

In $\widetilde{O}_1$ the matrix $T_1(y)$ has the form

$$T_1(y) = \begin{pmatrix} V_1(z) & W(z) \\ 0 & V_2(z) \end{pmatrix} (z - a_1)^{A_1}(y - a_1)^{E_1}, \tag{3.9}$$

where $V_1(z)$ and $V_2(z)$ are holomorphically invertible at a_1, and $W(z)$ is in general only meromorphic there. By the second condition in (3.2) and condition iii) of Lemma 5.3.1, the diagonal elements φ_1^j of A_1 satisfy the inequalities $\varphi_1^1 \ge \dots \ge \varphi_1^t$

and $\varphi_1^{t+1} \geq \cdots \geq \varphi_1^p$. Since E_1 has the same block form as G_1 in (3.1), we get that the matrix $L(z) = A_1 + (z-a_1)^{A_1} E_1 (z-a_1)^{-A_1}$ is holomorphic at a_1.

Denote by p the order of the pole of $W(z)$ at a_1. Then $W(z) = (z-a_1)^{-pI} W'$, where W' is holomorphic at a_1. Let us apply Lemma 4.1.1 to the matrix

$$(z-a_1)^C \begin{pmatrix} V_2(z) \\ W'(z) \end{pmatrix},$$

where $C = \operatorname{diag}(0, \dots, 0, -p, \dots, -p)$. We get

$$\Gamma'(z) \begin{pmatrix} V_2 \\ W \end{pmatrix} = \begin{pmatrix} V_2 \\ W'' \end{pmatrix}$$

with a holomorphic matrix W''. Let

$$S \begin{pmatrix} V_2 \\ W'' \end{pmatrix} = \begin{pmatrix} W'' \\ V_2 \end{pmatrix}, \qquad \widetilde{\Gamma} = S\Gamma' S^{-1}.$$

Then the system with fundamental matrix $\widetilde{T}(y) = \widetilde{\Gamma} T'(y)$ has the holomorphic factor

$$\begin{pmatrix} V_1 & W'' \\ 0 & V_2 \end{pmatrix}$$

in the corresponding decomposition (1.2.6). Therefore, the system with this fundamental matrix is Fuchsian at a_1.

REMARK 5.3.1. The assertion of the theorem does not depend on the arrangement of the irreducible representation χ_1 in (3.6). If the collection $G_1^2, \dots, G_n^2$ determines an irreducible representation of dimension l, and the matrix G_i' in (3.1) has order greater than $p-l-1$, then we get the same result for χ (under the assumption that χ_1 is realized as the monodromy representation of some Fuchsian system).

The proof is exactly the same as in the preceding theorem. It is only necessary to replace (3.7) by the conditions

$$\min_{\substack{i=2,\dots,n;\\ j=1,\dots,p-l}} \varphi_i^j = d_1, \qquad \min_{j=1,\dots,p-l} \varphi_1^j = d_2. \tag{3.10}$$

COROLLARY 5.3.1. *Suppose that all the monodromy matrices $\chi(\sigma_i)$ of the representation χ reduce simultaneously to the form (3.6), where the order of each G_j^1 is l. Assume that for some i*

$$G_i = \begin{pmatrix} G_i^1 & 0 \\ 0 & G_i^2 \end{pmatrix}.$$

If both the representations χ_1 and χ_2 with matrices $\chi_i(\sigma_j) = G_j^i$, $i = 1, 2$, can be realized as the monodromy matrices of some Fuchsian system, then the same holds for the representation χ.

The proof is a simplified version of the proof of Theorem 5.3.1. It is only necessary to make the following changes: omit all the conditions connected with d_2, let $t = l = \dim \chi_1$, and choose a Fuchsian system with the monodromy χ_1 in such a way that its normalizations φ_j^k for $j \neq i$ satisfy the condition $\varphi_j^k > d_1$, $k = 1, \dots, l$. The last condition is easily achieved by means of the substitution

$$\widetilde{T}^1(y) = \left(\prod_{j \neq i} (z-a_j)^d \right) (z-a_i)^{-d(n-1)} T^1(y)$$

with $d = d_1 + 1 - \min_{j\neq i;k=1,\dots,l}\varphi_j^k$. (We remark that the matrix of this transformation is holomorphic at the point ∞.)

The next assertion is a direct consequence of Theorem 5.3.1 and Lemma 5.4.1.

THEOREM 5.3.2. *Suppose that all the matrices* $\chi(\sigma_i)$ *reduce simultaneously to the form* (1.10), *where each collection* $G_1^j, \dots, G_n^j$ *forms an irreducible representation* χ_j. *If for each* j *there exists an index* i *such that the matrix* G_i *can be reduced by an upper triangular matrix* S_i^j *to a form* $G_i' = S_i^j G_i (S_i^j)^{-1}$ *with*

$$G_i' = \left(\begin{array}{ccc|ccc} (G_i^1)' & & * & & & \\ & \ddots & & & 0 & \\ 0 & & (G_i^j)' & & & \\ \hline & & & (G_i^{j+1})' & & * \\ & 0 & & & \ddots & \\ & & & 0 & & (G_i^k)' \end{array}\right), \tag{3.11}$$

then the representation χ *can be realized as the monodromy representation of some Fuchsian system.*

A negative solution of the RH problem means that there exist regular systems on the Riemann sphere that are not globally meromorphically equivalent to Fuchsian systems.

THEOREM 5.3.3. *Any system* (1.1.1) *with regular singular points is a subsystem (factor system) of some system with the same singular points that is meromorphically equivalent to a Fuchsian system.*

Theorem 5.3.3 is equivalent to the following assertion.

THEOREM 5.3.4. *Any representation* χ *is a subrepresentation (factor representation) of a representation for which the Riemann–Hilbert problem has a positive solution.*

Suppose that the representation χ of a system (1.1.1) with regular singular points $a_1, \dots, a_n$ is reducible. We consider a fundamental matrix $T(y)$ of the solution space of this system such that in the basis of its columns the monodromy matrices $\chi(\sigma_i)$ have the form

$$\chi_l(\sigma_i) = \begin{pmatrix} G_i^1 & & & * \\ 0 & G_i^2 & & \\ \vdots & & \ddots & \\ 0 & 0 & \cdots & G_i^m \end{pmatrix}. \tag{3.12}$$

where the collection of matrices $G_1^j, \dots, G_n^j$ is irreducible for all $j = 1, \dots, m$.

Suppose that the matrix G_i^l has order k_l. Denote by X_l the subspace of the solution space X of (1.1.1) generated by the first $k_1 + \dots + k_l$ columns of $T(z)$, and by χ^l the restriction of χ to X_l. The monodromy matrices of the subrepresentation χ^l are formed by the first $k_1 + \dots + k_l$ rows and columns of the matrices $\chi(\sigma_i)$ in (3.12).

LEMMA 5.3.2. *Suppose that the first element* $y_1(z)$ *of some row* $y_1, \dots, y_p$ *of the matrix* $T(y)$ *is nonzero. If any subrepresentation of the monodromy representation* χ *of the system* (1.1.1) *is indecomposable, then the functions* $y_1, \dots, y_p$ *are linearly independent.*

PROOF. Assume the contrary. Then there is a nonsingular upper triangular matrix S such that the elements of the row vector $(x_1, \dots, x_p) = (y_1, \dots, y_p)S$ with indices $j \in J$ are linearly independent, while the elements with indices $j \in \{1, \dots, p\} \setminus J = J^0$ are identically equal to zero. We consider the linear subspace Y of vector-valued functions generated by the columns t'_j of the matrix TS, $j \in J^0$.

For an arbitrary monodromy operator g^*,

$$g^*(t'_j) = \sum_{l=1}^{p} \lambda_l t'_l. \tag{3.13}$$

For the row of $T(y)S$ under consideration we get from (3.13) that for $t'_j \in Y$

$$0 \equiv g^*(x_j) = \sum_{l=1}^{p} \lambda_l x_l = \sum_{l \in J} \lambda_l x_l,$$

from which, since the elements x_l, $l \in J$, are linearly independent, we have that $\lambda_l = 0$ for $l \in J$. Therefore, the right-hand side of (3.13) is a linear combination of elements of the space Y. In other words, Y is an invariant subspace for all the monodromy operators.

By assumption, Y does not contain the first column of $T(y)$; therefore, $X_1 \not\subset Y$, and hence $X_1 \cap Y = \varnothing$, for otherwise χ^1 would be reducible. Let s be the first index for which $X_s \cap Y \neq \varnothing$. Then $X_{s-1} \cap Y = \varnothing$, and the irreducibility of the factor representation χ^s/χ^{s-1} gives us that all the columns of TS with indices $k_1 + \dots + k_{s-1} + 1, \dots, k_1 + \dots + k_s$ belong to Y; that is, $X_s \cap Y$ is an invariant subspace for χ^s and $\chi^s = \chi^{s-1} \oplus \chi^s/\chi^{s-1}$, which contradicts the indecomposability of the subrepresentation χ^s of χ. The lemma is proved.

REMARK 5.3.2. Lemma 5.3.2 implies, in particular, Lemma 1.3.4 on the linear independence of the elements of any row of a fundamental matrix of a system (1.1.1) with irreducible monodromy.

PROPOSITION 5.3.1. *Suppose that all the subrepresentations of a representation* χ *are indecomposable. If the monodromy matrices of this representation can be reduced simultaneously to the form* (3.12) *in such a way that for some* i *the matrix* $\chi(\sigma_i)$ *has block form*

$$\chi(\sigma_i) = \begin{pmatrix} G'_i & 0 \\ 0 & G''_i \end{pmatrix}, \tag{3.14}$$

where the order of G'_i *is at most* k_1, *then the Riemann–Hilbert problem for the representation* χ *has a positive solution.*

PROOF. It can be assumed without loss of generality that the matrix $\chi(\sigma_1)$ has the required form (3.12) and that it is upper triangular. We assume that $a_1 = 0$ (this can always be achieved by a conformal transformation of the Riemann sphere). If the block G'_i has order l, then $l \leq k_1$ by assumption.

We consider a system with regular singular points $a_1, \dots, a_n$ that is Fuchsian at $a_2, \dots, a_n$, together with a fundamental matrix $T(y)$ of the form (3.12) for the solution space of this system (the existence of a system with such a matrix $T(y)$ was proved in Corollary 5.1.1) such that the matrix $\chi(\sigma_1)$ has the form (3.12) in the basis of its columns. Suppose that in a neighborhood of the point $a_1 = 0$

$$T(y) = U_1(z)y^{E_1}; \tag{3.15}$$

then the matrix $U_1(z)$, which is meromorphic at zero, also has the form (3.12) with G_1^k replaced by U_1^k. We denote by r the order of the zero (or pole, if the sign is "$-$") of the function $\det U_1(z)$ at 0, and by r_1 the order of the zero (or pole, if the sign is "$-$") of the function $\det U_1(z)/\det U_1^1(z)$. Let $d_1, d_2, d_3, b_1, \dots, b_p$ be integers for which

$$\begin{gathered} d_1 > b_1 > \cdots > b_l, \qquad b_{l+1} > \cdots > b_p > d_2, \qquad b_j - b_{j+1} > d_3, \quad j \neq l, \\ r - ld_1 - (p-l)b_{l+1} > 0, \\ r_1 - (p-k_1)d_2 + \frac{(n-2)(p-k_1)(p-k_1-1)}{2} < 0. \end{gathered} \tag{3.16}$$

We represent the matrix $T(y)$ in (3.15) in the form

$$T(y) = \widetilde{U}_1(z)z^B y^{E_1}, \tag{3.16$'$}$$

where $B = \operatorname{diag}(b_1, \dots, b_p)$ and the b_j are from (3.16). Let us consider the corresponding matrix $\Gamma(z)$ and the decomposition (1.3.26) for the matrix $\widehat{U}_1(z)$.

LEMMA 5.3.3. *The inequality in Remark* 4.6.3 *holds under the conditions of Proposition* 5.3.1 *for the elements of the matrix* C *in* (1.3.26).

PROOF. We prove the lemma by induction on the number m of blocks in (3.12). For $m = 1$ the lemma follows from Theorem 4.2.1. Suppose it has been proved for $m = t$. We prove it for $m = t + 1$.

It follows from (3.16$'$) and (1.3.26) that $\sum i = 1^p c_i = r - \sum_{i=1}^p b_i$; therefore, by (3.16),

$$c_1 \geq \frac{1}{p}\sum_{i=1}^p c_i \geq \frac{1}{p}\left(r - \sum_{i=1}^l b_i - \sum_{i=l+1}^p b_i\right) \geq \frac{1}{p}(r - ld_1 - (p-l)b_{l+1}) > 0. \tag{3.17}$$

Let us consider the first row $(y_1, \dots, y_p)$ of $\Gamma(z)T(y)$.

LEMMA 5.3.4. *The element* $y_1(y)$ *of the first row of* $\Gamma(z)T(y)$ *is not identically equal to zero.*

PROOF. Assume that $y_1(y) \equiv 0$. Then $y_2(y) \equiv \cdots \equiv y_{k_1}(y) \equiv 0$ because the subrepresentation χ^1 is irreducible. Since $T(y)$ has the form (3.12), the latter means that the first k_1 elements of the first row γ of $\Gamma(z)$ are equal to 0. Since all the elements are holomorphic away from 0, it follows that

$$\gamma = (\underbrace{0, \dots, 0}_{k_1}, \gamma_1, \dots, \gamma_{p-k_1}),$$

and there are two possibilities: a) $\gamma_j = s \neq 0$, $s \in C$, for some j, $1 \leq j \leq p - k_1$; b) all the nonzero γ_j have poles at 0.

Consider first the case a). Let us take the matrix

$$\widehat{\Gamma}(z) = \left(\begin{array}{c|c|c} \gamma_1 & s & \gamma_{p-k_1} \\ I^{j-1} & 0 & \mathbf{0} \\ \hline & \vdots & \\ \mathbf{0} & 0 & I^{p-k_1-j} \end{array}\right)$$

and the matrix $T'(y)$ obtained from $T(y)$ by omitting the first k_1 rows and columns. Denote by U_1', B', and E_1' the matrices obtained by omitting the first k_1 rows and columns of the matrices $\widetilde{U}_1$, B, and E_1, respectively. We get from (1.3.26) that

$$\widehat{\Gamma}(z)U_1'(z) = \widehat{V}(z), \tag{3.18}$$

where the first row $\widehat{v}(z)$ of the matrix $\widehat{V}(z)$ has the form $\widehat{v}(z) = z^{c_1}v(z)$, and $v(z)$ is a row vector that is holomorphic at 0.

The matrix $T'(y)$ has a decomposition of the form (3.16$'$) at 0:

$$T'(y) = \widehat{U}_1'(z)z^{B'}y^{E_1'}. \tag{3.19}$$

For it we consider the corresponding matrix $\Gamma'(z)$ and a decomposition of the form (1.3.26) with matrices C' and $V'(z)$:

$$\Gamma'(z)\widehat{U}_1'(z) = z^{C'}V'(z). \tag{3.20}$$

According to the induction hypothesis,

$$\sum_{i=1}^{p-k_1}(c_1' - c_i') \le \frac{(n-2)(p-k_1)(p-k_1-1)}{2}. \tag{3.21}$$

On the other hand, by the second inequality in (3.16),

$$\sum_{i=1}^{p-k_1} c_i' = r_1 - (b_{k_1+1} + \cdots + b_p) < r_1 - (p-k_1)d_2, \tag{3.22}$$

where r_1 is the order of the zero of the function $\det U_1/\det U_1^1 = \det \widehat{U}_1'$ at 0. Combining (3.21) and (3.22) and using the last inequality in (3.16), we find that $(p-k_1)c_1' < 0$, and so

$$c_1' < 0. \tag{3.23}$$

Consider the matrix

$$\mathbb{Z}(z) = \widehat{\Gamma}(z)T'(y)(\Gamma'(z)T'(y))^{-1}.$$

On the one hand, $\mathbb{Z}(z)$ is meromorphic and holomorphically invertible away from the point $a_1 = 0$. On the other hand, for its first row $z_1(z)$ we get from (3.19), (3.20), and (3.23) that in a neighborhood of $a_1 = 0$

$$z_1(z) = z^{c_1 - c_1'}v(z)(V'(z))^{-1},$$

where the row vector $v(z)$ is holomorphic at zero. Since $v(z)$ is holomorphic, by (3.17) and (3.23) we have $z_1(0) = 0$. Since the row $z_1(z)$ is holomorphic on the whole Riemann sphere away from $a_1 = 0$, Liouville's theorem gives us that $z_1(z) \equiv 0$, which contradicts the holomorphic invertibility of $\mathbb{Z}(z)$ away from 0.

We consider case b). Denote by Γ_1^T the transpose of the matrix Γ_1 and by c_{k_1+i} the order of the pole of $\gamma_i(z)$ at zero. Suppose that $c_{k_1+j} = \min_i(c_{k_1+i})$. We apply the $A(0, k_1+j, 1, -c_{k_1+j})$ procedure of Lemma 1.3.2 to Γ_1^T and get a matrix ${}^1\Gamma$ such that the elements of its first column have poles of orders ${}^1c_{k_1+i}$ at zero. Suppose that $\min_i({}^1c_{k_1+i}) = {}^1c_{k_1+j_1}$. Then ${}^1c_{k_1+j_1} < c_{k_1+j}$. We apply the $A(0, k_1+j_1, 1-{}^1c_{k_1+j_1})$ procedure to ${}^1\Gamma$, and so on. In finitely many steps we get a matrix ${}^0\Gamma = \Gamma_0(z)\Gamma_1^T$ with first column holomorphic at 0. (Indeed, if in some step we get ${}^s\gamma_i(z) \equiv 0$ for $i \neq j$, for the elements of the first column of ${}^s\Gamma$, then ${}^s\gamma_j(z) \equiv \text{const}$, for if ${}^s\gamma_j(z)$ has a pole at zero, then ${}^s\gamma(z') = 0$ for some $z' \neq 0$, and then $\det {}^s\Gamma(z') = 0$, which contradicts the holomorphic invertibility of Γ_1^T away from 0.) Therefore, the matrix ${}^0\Gamma^T$ has first row satisfying condition a). The proof of b) now repeats in entirety that of a), with Γ_1 replaced by ${}^0\Gamma^T$, $T(y)$ by $(\Gamma_0^T)^{-1}T(y)$, and U_1' by $(\Gamma_0^T)^{-1}U_1'$. Lemma 5.3.4 is proved.

We return to the proof of Lemma 5.3.3. Consider again the first row of the matrix $\Gamma(z)T(y)$. According to Lemmas 5.3.2 and 5.3.4, the elements of this row are analytic linearly independent functions. The rest of the proof of Lemma 5.3.3 repeats in entirety the proof of necessity in Theorem 4.2.1. The only thing that must be checked is the holomorphicity of $z^B E_1 z^{-B}$. But it follows from the block form (3.12) of the matrix $E_1 = (2\pi i)^{-1}\ln\chi(\sigma_1)$ that this condition must be verified for each block separately, and the latter follows from the first inequalities in (3.16). Lemma 5.3.3 is proved.

We return to the proof of Proposition 5.3.1. According to Lemma 5.3.3, there is a matrix $\Gamma(z)$ that is meromorphic and holomorphically invertible away from the point $a_1 = 0$ such that

$$\Gamma(z)T(y) = z^C V(z) z^B y^{E_1}, \tag{3.24}$$

where $C = \operatorname{diag}(c_1, \dots, c_p)$, $c_1 \geq \dots \geq c_p$, the numbers c_i satisfy (4.2.1), $V(z)$ is holomorphically invertible, and B is a diagonal integer matrix whose elements satisfy the first three inequalities in (3.16). We choose the number d_3 in (3.16) so that

$$d_3 > (n-2)p(p-1)/2. \tag{3.25}$$

According to Lemma 4.1.3, there is a matrix $\Gamma_1(z)$ that is meromorphic on the Riemann sphere, holomorphically invertible away from $a_1 = 0$, and such that

$$\Gamma_1(z) z^C V(z) = V_1(z) z^{\widehat{C}} \tag{3.26}$$

where $\widehat{C}$ is a matrix obtained by a permutation of the diagonal elements of C, and $V_1(z)$ is a holomorphically invertible matrix.

We pass from the system with fundamental matrix $T(y)$ to a system with fundamental matrix $T_1(y) = \Gamma_1(z)\Gamma(z)T(y)$. Since $\Gamma_1(z)$ and $\Gamma(z)$ are holomorphically invertible away from $a_1 = 0$, the new system of equations will again be Fuchsian away from $a_1 = 0$. By (3.24) and (3.26), a fundamental matrix of it in a neighborhood of $a_1 = 0$ will have the form

$$T_1(y) = V_1(z) z^{B+\widehat{C}} y^{E_1}. \tag{3.27}$$

It follows from (4.2.1), (3.25), and (3.16) that the elements $\widehat{b}_1, \dots, \widehat{b}_p$ of the matrix $\widehat{B} = B + \widehat{C}$ satisfy the inequalities

$$\widehat{b}_1 > \cdots > \widehat{b}_l, \qquad \widehat{b}_{l+1} > \cdots > \widehat{b}_p. \tag{3.28}$$

In view of the block form (3.12), the upper triangularity of E_1, and (3.28) we again get that the matrix $z^{\widehat{B}} E_1 z^{-\widehat{B}}$ is holomorphic at zero. Consequently, by Corollary 1.2.2, the system constructed is Fuchsian at 0. Proposition 5.3.1 is proved.

PROPOSITION 5.3.2. *Suppose that all the factor representations of a representation χ are indecomposable. If the monodromy matrices of this representation can be reduced simultaneously to the form* (3.12) *in such a way that for some i the matrix $\chi(\sigma_i)$ has the form* (3.14), *where the order of G_i'' does not exceed k_m, then the Riemann–Hilbert problem for χ has a positive solution.*

PROOF. As in Proposition 5.3.1, we again assume that $i = 1$ and $a_1 = 0$, and we again consider a system with regular singular points $a_1, \dots, a_n$ that is Fuchsian at $a_2, \dots, a_n$ and has a fundamental matrix $T(y)$ of the form (3.12) for its solution space, with the matrix $\chi(\sigma_1)$ upper triangular and of the form (3.14) in the basis of columns of $T(y)$. We consider the decomposition (3.15) for this matrix.

Let $d_1, d_2, \varkappa, b_1, \dots, b_p$ be integers for which

$$\begin{aligned} b_i &= d_1 - \varkappa(i-1), && i = 1, \dots, l, \\ b_j &= d_2 + \varkappa(p-j), && j = l+1, \dots, p, \quad \varkappa > 0, \end{aligned} \tag{3.30}$$

where l is the order of G_1', $l \geq p - k_m$.

We represent the $T(y)$ in (3.15) in the form (3.16′), where $B = \operatorname{diag}(b_1, \dots, b_p)$, and the b_i are from (3.30); and we consider the corresponding matrix $\Gamma(z)$ in the decomposition (1.3.26) for the matrix $\widehat{U}_1(z)$. Denote by $(y_1, \dots, y_p)$ the first row of the matrix ΓT. Just as in Lemma 5.3.2 we consider a nonsingular upper triangular matrix S such that the elements of the row vector $(x_1, \dots, x_p) = (y_1, \dots, y_p)S$ with indices $j \in J$ are linearly independent, while the elements with indices $j \in J^0 = \{1, \dots, p\} \setminus J$ are identically equal to zero. By using the irreducibility of the representations χ^s/χ^{s-1} it can be proved just as in Lemma 5.3.2 that $J^0 = J^0_{m_1} \cup \cdots \cup J^0_{m_t}$, where $J^0_{m_i}$ is the set of indices of the columns of some block G_i^s in (3.12). The monodromy matrices of the fundamental matrix $\Gamma(z)T(y)$ are the same as for $T(y)$, and hence we get from the form of the row $x = (x_1, \dots, x_p)$ that the elements of the matrices $G_i' = S^{-1} G_i S$ whose indices (k, s) have $k \in J$ and $s \in J^0$ are equal to 0. Indeed, the row x passes into xG_i' under analytic continuation along the path g_i^{-1}. For $s \in J^0$ we get

$$0 \equiv g_i^*(x_s) = x_s(g_i^{-1} z) = \sum_{k \in J} g_{ks} x_k,$$

and so for the elements of the matrix G_i' we have $g_{ks} = 0$ for $k \in J$ and $s \in J^0$, because the x_k with $k \in J$ are linearly independent.

Denote by U_1', B', V', H_i', and A_i' the matrices obtained by omitting the rows and columns of the respective matrices $\widehat{U}_1 z^B S z^{-B}$, B, V, G_i', and A_i with indices in J^0. We remark that the indices of the columns of the last block G_i^m in

(3.12) do not appear in J^0, for otherwise the factor representation χ/χ^{s-1} would be decomposable, where G_i^s is the last block with the indices of its columns in J. Therefore, the matrix S has the form

$$S = \begin{pmatrix} \widehat{S} & 0 \\ 0 & I^{k_m} \end{pmatrix}, \tag{3.31}$$

where $\widehat{S}$ is an upper triangular matrix. From this and (3.30) we get that the matrix $z^B S z^{-B}$ is holomorphic at $z = 0$ and $\det(z^B S z^{-B}) = \det S \neq 0$. Consequently, V' is a holomorphically invertible matrix.

The row $(x_{i_1}, \dots, x_{i_q})$, $i_j \in J$, $q = |J|$, is analytic away from the points $a_1, \dots, a_n$, and has monodromy $H_1', \dots, H_n'$ and decompositions of the form

$$(x_{i_1}, \dots, x_{i_q}) = (u_1^i, \dots, u_q^i)(z - a_i)^{A_i'}(y - a_i)^{E_i'} S_i, \qquad i \neq 1,$$
$$(x_{i_1}, \dots, x_{i_q}) = z^{c_1}(v_1', \dots, v_q')z^{B'} y^{E_1'} S_1,$$

where $u_j^i(z)$ and $v_j'(z)$ are holomorphic at a_i and at $a_1 = 0$, respectively, and $E_j' = (2\pi i)^{-1} \ln H_i'$. (These decompositions follow from (1.2.26) and (3.24).)

Applying the procedure of Theorem 3.2.1 to this row, we get an equality of the type (3.1.27):

$$qc_1 + \sum_{j \in J} b_j + \sum_{i=2}^{n} \sum_{j \in J} \varphi_i^j + \sum_{i=1}^{n} \sum_{j \in J} \rho_i^j + d = \frac{(n-2)q(q-1)}{2}, \tag{3.32}$$

where $\rho_i^j \geq 0$ are the diagonal elements of the matrix E_i, and $d \geq 0$. For any preassigned value of the difference $d_1 - d_2$ we can always choose d_1 and d_2 in (3.30) so that the sum of the third, fourth, and fifth terms on the left-hand side of (3.32) is nonnegative. Therefore, from (3.32),

$$pc_1 \leq \frac{(n-2)p(p-1)}{2} - \frac{p}{q} \sum_{j \in J} b_j. \tag{3.33}$$

The equality (3.1.27) holds for the form $\omega = dT \cdot T^{-1}$. Computing it from (3.33), we get

$$\sum_{i=1}^{p} (c_1 - c_i) \leq K + \sum_{i=1}^{p} b_i - \frac{p}{q} \sum_{j \in J} b_j, \tag{3.34}$$

where K is a constant independent of B.

Let us estimate the difference $\mathscr{D} = \sum_{i=1}^{p} b_i - (p/q) \sum_{j \in J} b_j$. From the fact that the indices from $p - l + 1$ to p lie in J and from (3.30) we have

$$\mathscr{D} = \sum_{i=1}^{l} b_i - \frac{p}{q} \sum_{j \in \{1, \dots, l\} \cap J} b_j - \frac{p-q}{q} \sum_{j=l+1}^{p} b_j$$
$$\leq \sum_{i=1}^{l} b_i - \frac{p}{q} \sum_{j=l-m_1+1}^{l} b_j - \frac{p-q}{q} \sum_{j=l+1}^{p} b_j,$$

where $m_1 = |\{1, \dots, l\} \cap J|$. Again using (3.30), after simple computations we get

$$\mathscr{D} \le \left(l - \frac{m_1 p}{q}\right)(d_1 - d_2 + \varkappa) + \varkappa\left(\frac{p(p-q)}{2} - p(l - m_1)\left(1 - \frac{m_1}{q}\right)\right). \tag{3.35}$$

Since $q = p - l + m_1$, it follows that

$$\mathscr{D} \le \frac{(l - m_1)(p - l)}{p - l + m_1}(d_1 - d_2 + \varkappa(1 - p)) + \varkappa\frac{p(p - q)}{2}. \tag{3.36}$$

Suppose now that

$$\varkappa > p|K|. \tag{3.37}$$

We choose d_2 and d_1 so that the number $\mathscr{D}$ in (3.36) does not exceed 0. (If $l = m_1$, then $\mathscr{D} = 0$, but if $l > m_1$, then $(l-m_1)(p-l)/(p-l+m_1) > (p-l)/p > 0$, and we must choose $d_1 < d_2$ such that $d_2 - d_1 > \varkappa p^2(p-q)/2(p-l)$.) We now get from (3.34) that

$$c_1 - c_p \le \sum_{i=1}^{p}(c_1 - c_i) \le |K| \le \frac{\varkappa}{p}. \tag{3.38}$$

As in Proposition 5.3.1, there is again a matrix $\Gamma_1(z)$ such that (3.26) holds for the matrices C and V in (3.24). We pass from the system with fundamental matrix $T(y)$ to a system with matrix $T_1 = \Gamma_1 \Gamma T$. The rest of the proof of Proposition 5.3.2 completely repeats the proof of Proposition 5.3.1 from the formula (3.27). It is only necessary to check that the elements of the matrix $B + \widehat{C}$ satisfy the inequalities (3.28). But these inequalities follow immediately from (3.30) and (3.38).

Proposition 5.3.2 is proved.

PROOF OF THEOREM 5.3.3. It suffices to show that each representation χ is a subrepresentation (factor representation) of some representation satisfying the conditions of Proposition 5.3.2 (Proposition 5.3.1).

We prove the second of these assertions (the first is proved similarly).

Suppose that the representation χ of the form (3.12) has r decomposable subrepresentations $\chi^{s_1}, \dots, \chi^{s_r}$, $s_1 < \dots < s_r$, $r \ge 1$. Let $s = s_1$.

Assume that G_2^1 and G_3^s have been reduced to Jordan normal form with the help of matrices R^1 and R^s (for this it suffices to pass to the matrices $\widehat{G}_i = R^{-1} G_i R$, where R has the form (3.12) with $R^k = I$ for $k \ne 1, s$), and that there are zeros above the blocks G_i^s in G_i (since χ^s is decomposable). Let $t_2 = g_{11}^1$ and $t_3 = g_{11}^s$, where $G_3^s = ((g_{ij}^s))$ and $G_2^1 = ((g_{ij}^1))$. Denote by b the vector

$$(\underbrace{1, 0, \dots, 0}_{k_1 + \dots + k_{s-1}}, 1, 0, \dots, 0).$$

Consider a representation χ of dimension $p + 1$ with matrices

$$\begin{gathered} G_1' = \begin{pmatrix} t_1 & 0 \\ 0 & G_1 \end{pmatrix}, \qquad G_2' = \left(\begin{array}{c|c} t_2 & b \\ \hline 0 & G_2 \end{array}\right), \qquad G_3' = \left(\begin{array}{c|c} t_3 & t_{11} \cdots t_{1p} \\ \hline 0 & G_3 \end{array}\right), \\ G_i' = \begin{pmatrix} 1 & 0 \\ 0 & G_i \end{pmatrix}, \qquad i > 3, \end{gathered} \tag{3.39}$$

where $t_{1j} = -(1/t_2)(\widehat{g}_{1j} + \widehat{g}_{lj})$, where $l = k_1 + \cdots + k_{s-1} + 1$, $t_1 = 1/(t_2 t_3)$, and $G_3 = ((\widehat{g}_{ij}))$. (We remark that $t_{ij} \neq 0$.)

This representation has at most $r-1$ decomposable subrepresentations, and the matrix G_1' has a form (3.14) satisfying the conditions of Proposition 5.3.1. Repeating the procedure, we get in $r-1$ steps a representation $\widehat{\chi}$ satisfying the conditions of Proposition 5.3.1 with the original representation χ as a factor representation.

If χ does not have decomposable subrepresentations, and the number m of blocks in (3.12) is greater than 1, then in the general case it is still necessary to go through the procedure once (for example, for $s = 2$) in order to get a matrix G_1' of the form (3.14). Here it suffices to reduce only the matrix G_2^1 to Jordan normal form. Theorem 5.3.3 is proved.

This also proves Theorem 5.3.2.

COROLLARY 5.3.2. *The dimension of the representation constructed does not exceed* $p + m + 1$, *where* m *is the number of irreducible blocks in the decompositon* (3.12) *for the monodromy matrices of the original representation* χ.

EXAMPLE 5.3.1. We consider a system with regular singular points $0, -1, 1, 1/2$:

(3.40)
$$df = \left(\begin{pmatrix} 0 & 0 & 0 & 0 \\ 0 & 0 & 1 & 0 \\ 0 & 0 & z & 0 \\ 0 & 0 & 0 & -z \end{pmatrix} \frac{dz}{z^2} + \frac{1}{6}\begin{pmatrix} 0 & 6 & 0 & 0 \\ 0 & 0 & 6 & 0 \\ 0 & 0 & -1 & 1 \\ 0 & 0 & -1 & 1 \end{pmatrix} \frac{dz}{z+1} + \frac{1}{2}\begin{pmatrix} 0 & 0 & 0 & 0 \\ 0 & 0 & 0 & 2 \\ 0 & 0 & -1 & -1 \\ 0 & 0 & 1 & 1 \end{pmatrix} \frac{dz}{z-1} + \frac{1}{3}\begin{pmatrix} 0 & -3 & 0 & 0 \\ 0 & 0 & -3 & -3 \\ 0 & 0 & -1 & 1 \\ 0 & 0 & -1 & 1 \end{pmatrix} \frac{dz}{z-1/2}\right) f.$$

The factor system of it obtained by omitting the first rows and columns of the coefficient matrices in (3.40) is not meromorphically equivalent to a Fuchsian system, since the Riemann–Hilbert problem has a negative solution (see (2.4.1)) for the monodromy representation of this factor system. But the system (3.40) itself, with the substitution $\Gamma_2(z)\Gamma_1(z)f = g$, where

$$\Gamma_1(z) = \begin{pmatrix} 1 & 0 & 0 & 0 \\ -\frac{1}{3z} & 1 & 0 & 0 \\ 0 & 0 & 1 & 0 \\ 0 & 0 & 0 & 1 \end{pmatrix}, \qquad \Gamma_2(z) = \begin{pmatrix} 1 & 0 & 0 & 0 \\ 0 & 1 & -\frac{1}{z} & 0 \\ 0 & 0 & 1 & 0 \\ 0 & 0 & 0 & 1 \end{pmatrix},$$

reduces to the Fuchsian system

(3.41)
$$dg = \left(\begin{pmatrix} 1 & 0 & 3 & 0 \\ -1/3 & -1 & -2 & 0 \\ 0 & 0 & 1 & 0 \\ 0 & 0 & 0 & -1 \end{pmatrix} \frac{dz}{z} + \frac{1}{6}\begin{pmatrix} -2 & 6 & -6 & 0 \\ 2/3 & 2 & 3 & 1 \\ 0 & 0 & -1 & 1 \\ 0 & 0 & -1 & 1 \end{pmatrix} \frac{dz}{z+1} + \frac{1}{2}\begin{pmatrix} 0 & 0 & 0 & 0 \\ 0 & 0 & 1 & 3 \\ 0 & 0 & -1 & -1 \\ 0 & 0 & 1 & 1 \end{pmatrix} \frac{dz}{z-1} + \frac{1}{3}\begin{pmatrix} -2 & -3 & -6 & 0 \\ 2/3 & 2 & 3 & -5 \\ 0 & 0 & -1 & 1 \\ 0 & 0 & -1 & 1 \end{pmatrix} \frac{dz}{z-1/2}\right) g.$$

EXAMPLE 5.3.3. Since any system of two equations with n regular singular points and any system of three equations with three regular singular points are meromorphically equivalent to Fuchsian systems (see Proposition 1.5.3 and Theorem 2.3.2), Example 5.3.1 is minimal in this sense (with respect to the number of equations of the system and the number of singular points).

REMARK 5.3.4. The system (3.41) is also the first (with respect to the number of equations and singular points) example of a Fuchsian system that cannot be reduced by a meromorphic substitution to a Fuchsian system with the same singular points whose coefficient matrices have the same character of reducibility (3.12) as its monodromy matrices. We note that the coefficient matrices of a regular system (3.40) have the required form (with three blocks of dimensions 1, 1, and 2, respectively). An example of a monodromy of a seventh-order system having the indicated property was given in Proposition 5.1.1.

EXAMPLE 5.3.2. Consider the matrices

$$G_1' = \begin{pmatrix} 1 & 1 & 0 & 0 & 0 \\ 0 & 1 & 1 & 0 & 0 \\ 0 & 0 & 1 & 1 & 1 \\ 0 & 0 & 0 & 1 & 1 \\ 0 & 0 & 0 & 0 & 1 \end{pmatrix}, \quad G_2' = \begin{pmatrix} 3 & 1 & 1 & -1 & 0 \\ -4 & -1 & 1 & 2 & 0 \\ 0 & 0 & 3 & 1 & 0 \\ 0 & 0 & -4 & -1 & -1 \\ 0 & 0 & 0 & 0 & 1 \end{pmatrix},$$

$$G_3' = \begin{pmatrix} -1 & 0 & 2 & -1 & 0 \\ 4 & -1 & 0 & 1 & 0 \\ 0 & 0 & -1 & 0 & 0 \\ 0 & 0 & 4 & -1 & 0 \\ 0 & 0 & 0 & 0 & 1 \end{pmatrix} \tag{3.42}$$

and an arbitrary triple a_1, a_2, a_3 of points.

A representation χ with matrices G_1', G_2', and G_3' satisfies the conditions of Proposition 5.3.2 (G_3' has the form (3.14)); therefore, the Riemann–Hilbert problem has a positive solution for it. At the same time, the four-dimensional subrepresentation of this representation with matrices formed by the first four rows and columns of the matrices (3.42) cannot be the monodromy representation of a Fuchsian system (see Example 5.2.1).

Bibliography

1. E. L. Ince, *Ordinary differential equations*, Longmans, Green, and Co., London, 1926; reprint, Dover, New York, 1944.
2. V. I. Arnol′d and Yu. S. Il′yashenko, *Ordinary differential equations*, Itogi Nauki i Tekhniki: Sovremennye Problemy Mat.: Fundamental′nye Napravleniya, vol. 1, VINITI, Moscow, 1985, pp. 7–149; English transl. in Encyclopaedia of Math. Sci., vol. 1 [Dynamical Systems, I], Springer-Verlag, Berlin, 1988.
3. A. A. Bolibrukh, *On the Riemann-Hilbert problem for systems of three equations*, Questions of Contemporary Mathematics in Physics and Mechanics Problems, Moskov. Fiz.-Tekhn. Inst., 1989, pp. 10–24. (Russian)
4. ———, *The Riemann-Hilbert problem on the complex projective line*, Mat. Zametki **46** (1989), no. 3, 118–120. (Russian)
5. ———, *The Riemann-Hilbert problem*, Uspekhi Mat. Nauk **45** (1990), no. 2 (272), 3–47; English transl. in Russian Math. Surveys **45** (1990).
6. ———, *Construction of a Fuchsian equation from a monodromy representation*, Mat. Zametki **48** (1990), no. 5, 22–34; English transl. in Math Notes **48** (1990).
7. ———, *On sufficient conditions for the positive solvability of the Riemann-Hilbert problem*, Mat. Zametki **51** (1992), no. 2, 9–19; English transl. in Math Notes **51** (1992).
8. Wolfgang Wasow, *Asyumptotic expansions for ordinary differential equations*, Interscience, New York, 1965.
9. N. P. Vekua, *Systems of singular integral equations and some boundary value problems*, 2nd rev. ed., "Nauka", Moscow, 1970; English transl. of 1st ed., Noordhoff, Groningen, 1967.
10. A. B. Venkov, *Examples of efficient solution of the Riemann-Hilbert problem on recovering a differential equation from the monodromy group in the framework of the theory of automorphic functions*, Zap. Nauchn. Sem. Leningrad. Otdel. Mat. Inst. Steklov (LOMI) **162** (1987), 5–42; English transl. in J. Soviet Math. **46** (1989), no. 2.
11. F. R. Gantmakher, *Theory of matrices*, 4th ed., "Nauka", Moscow, 1988; English transl. of 1st ed., Vols. 1, 2, Chelsea, New York, 1959.
12. V. V. Golubev, *Lectures on the analytic theory of differential equations*, GITTL, Moscow, 1950. (Russian)
13. V. A. Golubeva, *Some questions in the analytic theory of Feynman integrals*, Uspekhi Mat. Nauk **31** (1976), no. 2 (188), 135–202; English transl. in Russian Math. Surveys **31** (1976).
14. N. P. Erugin, *The Riemann problem*, "Nauka i Tekhnika", Minsk, 1982. (Russian)
15. P. G. Zograf and L. A. Takhtadzhyan, *On the Liouville equation, accessory parameters, and the geometry of Teichmüller space for Riemann surfaces of genus* 0, Mat. Sb. **132 (174)** (1987), 147–166; English transl. in Math. USSR Sb. **60** (1988).
16. Earl A. Coddington and Norman Levinson, *Theory of ordinary differential equations*, McGraw–Hill, New York, 1955.
17. B. L. Krylov, *Solution in finite form of the Riemann problem for a Gauss system*, Trudy Kazan. Aviatsion. Inst. **31** (1956), 203–445. (Russian)
18. I. A. Lappo-Danilevskiĭ, *Mémoires sur la théorie des systèmes des équations différentielles linéaires*, Trudy Fiz.-Mat. Inst. Steklov. Otdel. Mat. **6** (1934); **7** (1935); **8** (1936); reprint, Chelsea, New York, 1953.
19. N. I. Muskhelishvili, *Singular integral equations*, 3rd ed., "Nauka", Moscow, 1968; English transl. of 1st ed., Noordhoff, Groningen, 1953; reprinted 1972.

20. Christian Okonek, Michael Schneider, and Heinz Spindler, *Vector bundles on complex projective spaces*, Birkhäuser, Basel, 1980.
21. D. Hilbert, *Mathematische Probleme*, Lecture presented at the Second Internat. Congress Math. (Paris, 1900), Nachr. Ges. Wiss. Göttingen Math.-Phys. Kl. **1900**, 253–297; reprinted in Arch. Math. Phys. (3) **1** (1901), 44–63, 213–237, and also in his *Gesammelte Abhandlungen*. Vol. III, Springer-Verlag, Berlin, 1935; English transl., Bull. Amer. Math. Soc. **8** (1902), 437–479; reprinted in Mathematical Developments Arising from Hilbert Problems, Proc. Sympos. Pure Math., vol. 28, Amer. Math. Soc., Providence, RI, 1976, pp 1–34.
22. H. Poincaré, *Sur les groupes des équations linéaires*, Acta Math. **4** (1884), 201–312; reprinted in his *Œuvres*, Vol. II, Gauthier-Villars, Paris, 1916, pp. 300–401.
23. Bernhard Riemann, *Gesammelte mathematische Werke* (2nd ed.) and *Wissenschaftlicher Nachlass*, Teubner, Leipzig, 1892, 1910; combined reprint, Dover, New York, 1953.
24. M. Sato, M. Jimbo, and T. Miwa, *Holonomic quantum fields*. I, II, III, IV, V, and Supplement, Publ. Res. Inst. Math. Sci. **14** (1978), 223–254; **15** (1979), 201–278; **15** (1979), 577–629; **15** (1979), 871–972; **16** (1980), 531–584; **17** (1981), 137–151.
25. Otto Forster, *Riemannsche Flächen*, Springer-Verlag, Berlin and New York, 1977.
26. Philip Hartmann, *Ordinary differential equations*, Wiley, New York, 1964.
27. Dale Husemoller, *Fibre bundles*, McGraw–Hill, New York, 1966.
28. D. V. Anosov, *An introduction to Hilbert's twenty-first problem*, IMA Preprint Ser., no. 861, Minneapolis, 1991.
29. D. V. Anosov and A. A. Bolibruch [Bolibrukh], *The Riemann–Hilbert problem*, Vieweg, Braunschweig, 1994.
30. Kazuhiko Aomoto, *Une remarque sur la solution des équations de L. Schlesinger et Lappo-Danilevsky*, J. Fac. Sci. Univ. Tokyo Sect. I **17** (1970), 341–354.
31. M. F. Atiyah, *Complex analytic connections in fibre bundles*, Trans. Amer. Math. Soc. **85** (1957), 181–207.
32. W. Balser, *Analytic transformation to Birkhoff standard form in dimension three*, Funkcial. Ekvac. **33** (1990), 56–67.
33. A. A. Bolibruch [Bolibrukh], *Fuchsian systems with reducible monodromy and the Riemann–Hilbert problem*, Global Analysis—Studies and Applications, V, Lecture Notes in Math., vol. 1520, Springer-Verlag, Berlin and New York, 1992, pp. 139–155.
34. ———, *Fuchsian systems and holomorphic vector bundles on the Riemann sphere*, Preprint no. 347, Univ. Nice, Nice, 1993.
35. ———, *On analytic transformation to Birkhoff standard form*, Trudy Mat. Inst. Steklov. **203** (1994), 33–40; English transl. in Proc. Steklov Inst. Math. **1995**, no. 3 (203).
36. ———, *Hilbert's twenty-first problem for Fuchsian linear systems*, Developments in Mathematics: The Moscow School, Chapman and Hall, London, 1993, pp. 54–99.
37. G. D. Birkhoff, *A theorem on matrices of analytic functions*, Math. Ann. **74** (1913), 122–133.
38. ———, *Collected mathematical papers*. Vol. I, Amer. Math. Soc., Providence, RI, 1950, pp. 259–306.
39. B. Bojarski, *Connections between complex and global analysis: some analytical and geometrical aspects of the Riemann–Hilbert transmission problem*, Complex Analysis: Methods, Trends, and Applications (E. Lanckau and W. Tutschke, editors), Akademie-Verlag, Berlin, and North-Holland, Amsterdam, 1983, pp. 97–110.
40. Willibrordus Johannes Maria Dekkers, *Moduli spaces for pairs of* 2×2 *matrices and for certain connections on* $P^1(C)$, Dissertation, Catholic University of Nijmegen, Nijmegen, 1976, Krips Repro B. V., Meppel, 1976.
41. ———, *The matrix of a connection having regular singularities on a vector bundle of rank* 2 *on* $P^1(C)$, Équations Différentielles et Systèmes de Pfaff dans le Champ Complexe (Sém., Strasbourg, 1975), Lecture Notes in Math., vol. 712, Springer-Verlag, Berlin and New York, 1979, pp. 33–43.
42. P. Deligne, *Équations différentielles a points singuliers reguliers*, Lecture Notes in Math., vol. 163, Springer-Verlag, Berlin and New York, 1970.
43. A. Douady, *Problème de Riemann–Hilbert*. II: *Solution pour des points singuliers reels*, Mathématique et Physique (Sém., Paris, 1979/82; L. Boutet de Monvel et al., editors), Progr. Math., vol. 37, Birkhäuser, Boston, MA, 1983, pp. 289–298.
44. W. B. Jurkat, D. A. Lutz, and A. Peyerimhoff, *Birkhoff invariants and effective calculations for meromorphic differential equations*. I, II, J. Math. Anal. Appl. **53** (1976), 438–470; Houston J. Math. **2** (1976), 207–238.
45. N. M. Katz, *An overview of Deligne's work on Hilbert's twenty-first problem*, Mathematical Developments Arising from Hilbert Problems, Proc. Sympos. Pure Math., vol. 28, Amer. Math. Soc., Providence, RI, 1976, pp. 537–559.

46. A. Treibich-Kohn, *Un résultat de Plemelj*, Mathématique et Physique (Sém., Paris, 1979/82; L. Boutet de Monvel et al., editors), Progr. Math., vol. 37, Birkhäuser, Boston, MA, 1983, pp. 307–312.
47. V. P. Kostov, *Fuchsian systems on* CP^1 *and the Riemann-Hilbert problem*, Preprint No. 303, Univ. Nice, Nice, 1991.
48. ———, *Fuchsian linear systems on* CP^1 *and the Riemann-Hilbert problem*, C. R. Acad. Sci. Paris Sér. I Math. **315** (1992), 143–148.
49. A. H. M. Levelt, *Hypergeometric functions*, Nederl. Akad. Wetensch. Proc. Ser. A 64 = Indag. Math. **23** (1961), 361–403.
50. L. Boutet de Monvel, *Problème de Riemann-Hilbert*, Mathématique et Physique (Sém., Paris, 1979/82; L. Boutet de Monvel et al., editors), Progr. Math., vol. 37, Birkhäuser, Boston, MA, pp. 313–322.
51. M. Ohtsuki, *On the number of apparent singularities of a linear differential equation*, Tokyo J. Math. **5** (1982), 23–29.
52. J. Plemelj, *Problems in the sense of Riemann and Klein*, Interscience, New York, 1964.
53. H. Röhrl, *Das Riemann-Hilbertsche Problem der Theorie der linearen Differentialgleichungen*, Math. Ann. **133** (1957), 1–25.
54. ———, *Holomorphic vector bundles over Riemann surfaces*, Bull. Amer. Math. Soc. **68** (1962), 125–160.
55. T. Saito, *Problem of Riemann*, Sugaku **12** (1961), 145–159. (Japanese)
56. L. Schlesinger, *Über Lösungen gewisser Differentialgleichungen als Funktionen der singulären Punkte*, J. Reine Angew. Math. **129** (1905), 287–294.